Volume 1 of 2 Engine Maintenance Manual LD 465-1 / LD 465-1C / LT 465-1C LDS-465-1A / LDS 465-2 Engines TM 9-2815-210-34-2-1 for 2.5 Ton 6x6 M44A2 Truck Series aka Deuce and a Half

April 1981

AF808900

edited by
Brian Greul

The M44A2 series of military trucks is affectionately known as the deuce and a half, or simply the deuce. These ubiquitous trucks were first produced during WWII and General Eisenhower credited them as being one of the most important US Army vehicles. This book is volume 1 of 2 for the engine maintenance manual for the LD-465 series of diesel engines. It is being printed for enthusiasts, restores, and collectors who may wish to own a quality paperback copy. The editor has endeavored to minimize changes, but the following changes were made: Feedback forms are omitted, the fonts have been updated for printing purposes on modern equipment.

Should you have suggestions or feedback on ways to improve this book please send email to Books@OcotilloPress.com

---

Edited 2021 Ocotillo Press
ISBN 978-1-954285-47-7

Printed in the United States of America

Ocotillo Press
Houston, TX 77017
Books@OcotilloPress.com

TM 9-2815-210-34-2-1
T.O. 38G1-48-12-2-1

TECHNICAL MANUAL
VOLUME 2 OF 2
PART 1 OF 2

MAINTENANCE

DIRECT SUPPORT AND GENERAL SUPPORT LEVEL

ENGINE ASSEMBLY, DIESEL (MULTIFUEL): NATURALLY ASPIRATED OR TURBOCHARGED, FUEL-INJECTED, WATER-COOLED, 6-CYLINDER

MODELS: LD-465-1, NSN 2815-00-239-5824; LD-465-1C, NSN 2815-00-134-4830; LDT-465-1C, NSN 2815-00-103-2642; LDS-465-1, NSN 2815-00-075-0087; LDS-465-1A, NSN 2815-00-239-5819; AND LDS-465-2, NSN 2815-00-808-8011 AND CLUTCHES

NOTE:
THE STYLE OF THIS TM IS EXPERIMENTAL. IT IS BEING TRIED BY THE ARMY ONLY ON A LIMITED BASIS

DEPARTMENTS OF THE ARMY AND THE AIR FORCE
APRIL 1981

WARNING

SERIOUS OR FATAL INJURY TO PERSONNEL MAY RESULT IF THE FOLLOWING INSTRUCTIONS ARE NOT COMPLIED WITH.

Engine weighs 1600 pounds. Do not use a chain hoist and lifting sling of less than two-ton capacity. Engine could fall and cause injury to personnel.

Make sure lifting sling is pulled tight before taking mounting hardware out of engine mounting brackets in shipping container . Engine could slip off suspension frame and fall, causing injury to personnel .

Smoking, sparks or open flame are not allowed within 50 feet of work area when draining fuel . Fire or explosion could occur, causing injury to personnel and damage to equipment.

Pressure plate assembly is heavy. It must be held firmly up and against flywheel when mounting screws are taken out. If it is not held up, it will fall and cause injury to personnel .

Clutch disk is heavy enough to cause injury if dropped. It is held in place by pressure plate assembly. Do not let it fall out and drop when pressure plate assembly is taken out or it may cause injury to personnel.

Flywheel housing is heavy . Be sure to hold it firmly up and against engine when taking off or putting on mounting nuts. It could fall off mounting studs and cause injury to personnel .

Crankshaft damper and pulley assembly is heavy. Dropping it will cause damage to equipment and injury to personnel.

Intake manifold and exhaust manifold assembly is heavy. Hold it up when taking off locknuts and washers or it may fall and cause injury to personnel.

Starter assembly weighs fifty pounds . Be careful o hold it up firmly when taking it off mounting studs . Starter could fall and cause injury to personnel and damage to equipment.

* TM 9-2815-210-34-2-1
T.O. 38G1-48-12-2-1

TECHNICAL MANUAL
NO. 9-2815-210-34-2-1
TECHNICAL ORDER
NO. 38G1-48-12-2-1

DEPARTMENTS OF THE ARMY
AND
THE AIR FORCE
Washington DC, 3 April 1981

TECHNICAL MANUAL

VOLUME 2 OF 2

PART 1 OF 2

MAINTENANCE

DIRECT SUPPORT AND GENERAL SUPPORT LEVEL

ENGINE ASSEMBLY, DIESEL (MULTIFUEL): NATURALLY ASPIRATED OR TURBOCHARGED, FUEL-INJECTED, WATER-COOLED, 6-CYLINDER

MODELS: LD-465-1, NSN 2815-00-239-5824; LD-465-1C, NSN 2815-00-134-4830; LDT-465-1C, NSN 2815-00-103-2642; LDS-465-1, NSN 2815-00-075-0087; LDS-465-1A, NSN 2815-00-239-5819; AND LDS-465-2, NSN 2815-00-808-8011 AND CLUTCHES

REPORTING OF ERRORS AND RECOMMENDING IMPROVEMENTS

You can help improve this manual. If you find any mistakes or if you know of a way to improve the procedure, please let us know. Mail your letter, DA Form 2028 (Recommended Changes to Publications and Blank Forms).or DA Form 2028-2 located in the back of this manual direct to: Commander, US Army Tank-Automotive Command. ATTN: DRSTA-MB, Warren, Michigan 48090. A reply will be furnished to you.

TABLE OF CONTENTS

* This manual, together with TM 9-2815-210-34-1, 3 April 1981, and TM 9-2815-210-34-2-2, 3 April 1981, supersedes TM 9-2815-210-34, 16 June 1978.

TABLE OF CONTENTS-CONT

TABLE OF CONTENTS-CONT

TABLE OF CONTENTS-CONT

TABLE OF CONTENTS-CONT

TABLE OF CONTENTS-CONT

TABLE OF CONTENTS-CONT

Part 2 of 2
(TM 9-2815-210-34-2-2)

LIST OF ILLUSTRATIONS

LIST OF ILLUSTRATIONS - CONT

LIST OF TABLES

LIST OF TABLES - CONT

LIST OF TABLES - CONT

LIST OF TABLES - CONT

CHAPTER 1

GENERAL MAINTENANCE INFORMATION

1-1. SCOPE.

a. This manual gives instructions for direct support and general support maintenance of multifuel engine models LD-465-1, LD-465-1C, LDT-465-1C, LDS-465-1, LDS-465-1A, and LDS-465-2.

b. Engine models LD-465-1 and LDT-465-1C are used in the M44A2 series 2 1/2-ton trucks. Engine models LDS-465-1 and LDS-465-1A are used in the M39A2 series 5-ton trucks. The LDS-465-2 and LD-465-1C engines are used in the M656, M757, and M791 series 5-ton trucks.

c. This manual has procedures for disassembly, cleaning, inspection, repair, test, adjustment, and overhaul as authorized by the maintenance allocation chart.

d. Refer to TM 9-2815-210-34P for a listing of repair parts and special tools for the maintenance of multifuel engines covered in this manual.

e. Figures 1-1 to 1-11 show the multifuel engines covered in this manual.

1-2. GENERAL MAINTENANCE. General maintenance tasks relating to inspection, care and maintenance of antifriction bearings are given in TM 9-214. Welding procedures that apply to this type of equipment are given in TM 9-237. For lubrication, refer to LO 9-2320-209-12/1 and LO 9-2320-211-12.

1-3. CLEANING. Refer to TM 9-247 for general cleaning materials for this type of equipment. Special cleaning procedures are given in chapter 4.

1-4. PAINTING. For painting instructions for field use of the equipment covered in this manual, refer to TM 43-0139.

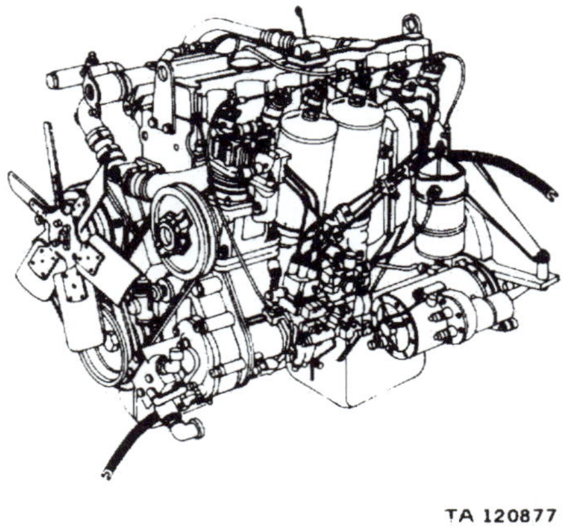

Figure 1-1. Model LDS-465-1 engine assembly, left front view

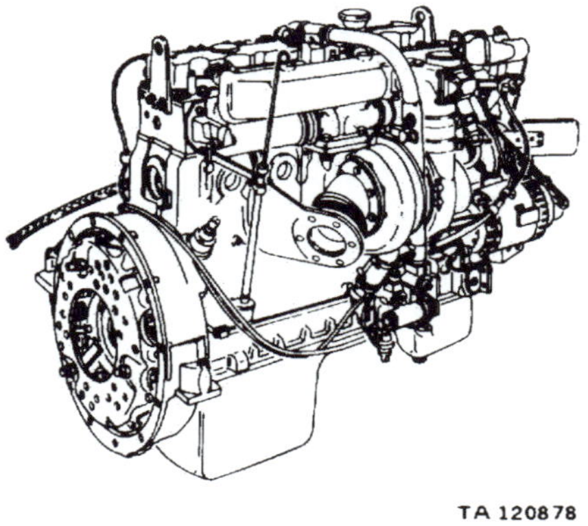

Figure 1-2. Model LDS-465-1 engine assembly, right rear view

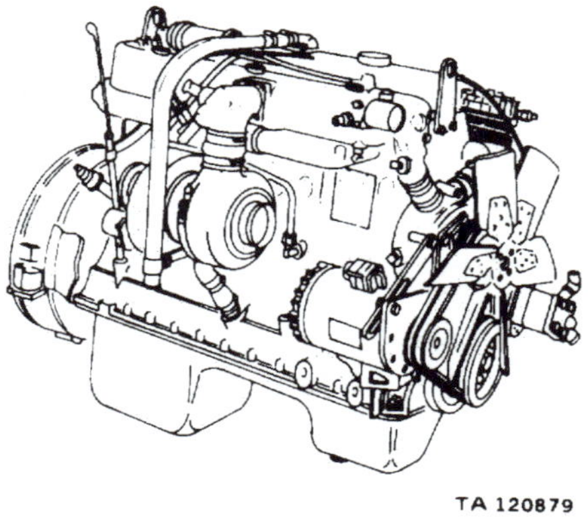

Figure 1-3 . Model LDS-465-1A engine assembly, right front view

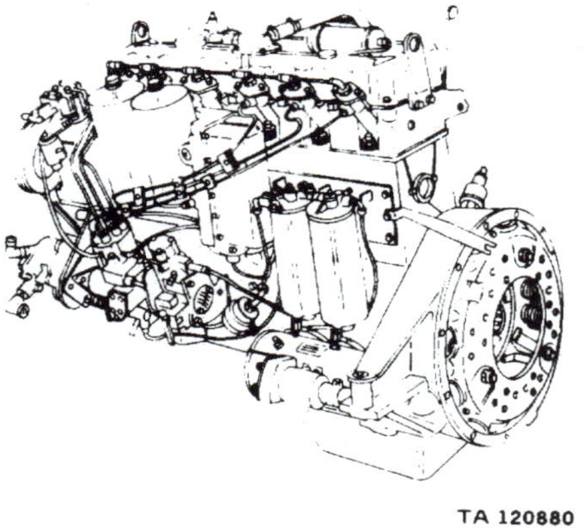

Figure 1-4. Model LDS-465-1A engine assembly, left rear view

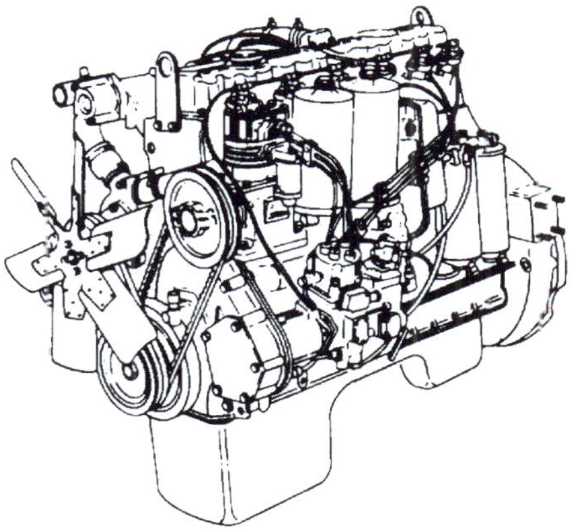

TA 120831

Figure 1-5. Model LD-465-1 engine assembly, left front view

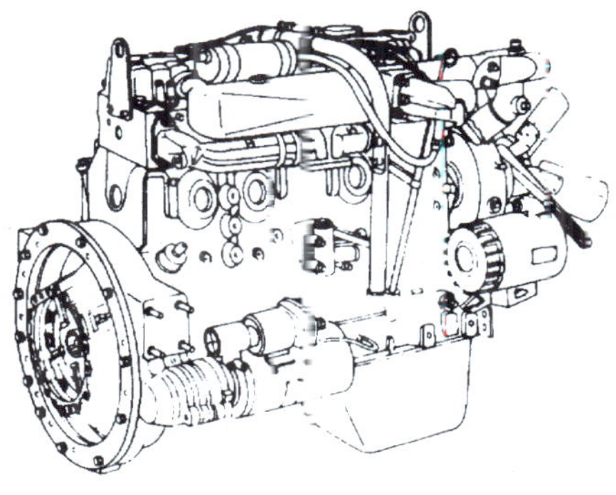

TA 120882

Figure 1-6. Models LD-465-1 engine assembly, right rear view

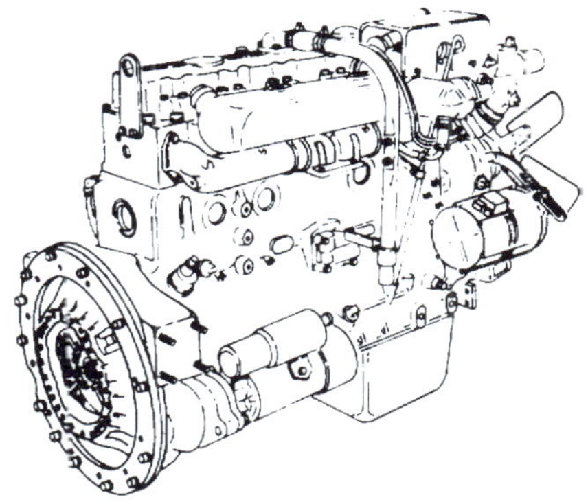

TA 120883

Figure 1-7. Model LD-465-1C engine assembly, right rear view

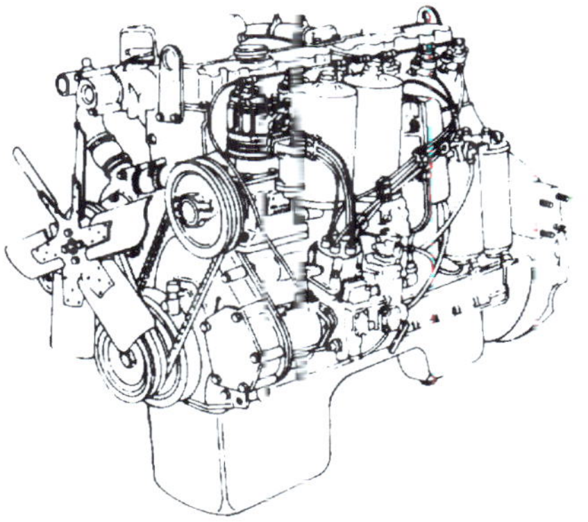

TA 120884

Figure 1-8. Models LD-465-1C and LDT-465-1C engine assemblies, left front view

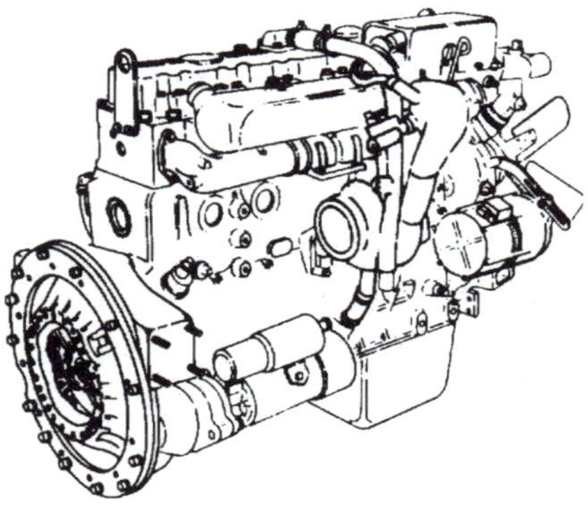

Figure 1-9. Model LDT-465-1C engine assembly, right rear view

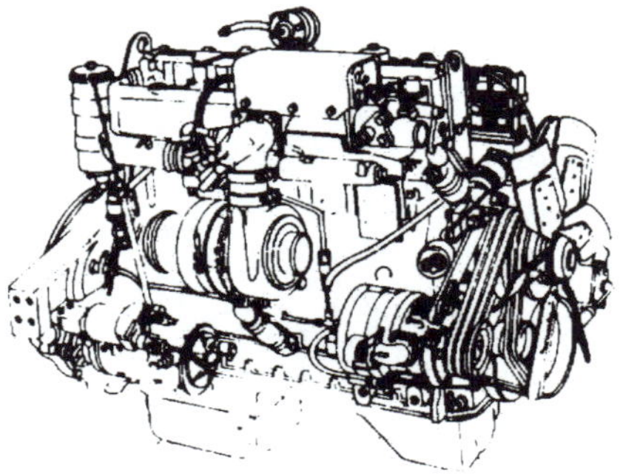

Figure 1-10. Model LDS-465-2 engine assembly, right front view

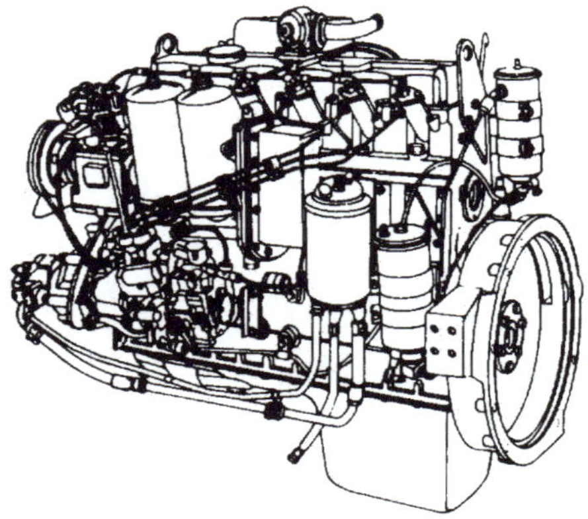

Figure 1-11. Model LDS-465-2 engine assembly, left rear view

1-5. TORQUE VALUES. Critical torque values for a particular component are given in the maintenance procedures. When torque values are not given bolts, screws and nuts are to be tightened as given in table 1-1.

1-6. SPECIAL TOOLS AND EQUIPMENT. Special tools and equipment are provided to make it easier to do particular maintenance tasks and to keep the equipment in good repair . Table 1-2 lists the special tools and equipment and gives reference to the maintenance paragraph where they are used and what they are used for.

1-7. SAFETY INSPECTION AND TESTING OF LIFTING DEVICES. Refer to TB 43-0142 for safety inspection and testing of lifting devices used in this manual.

1-8. FORMS AND RECORDS. Maintenance forms, records, and reports which are to be used by maintenance personnel at all levels are listed in and prescribed by TM 38-750.

1-9. EQUIPMENT IMPROVEMENT REPORT AND MAINTENANCE DIGEST (EIR MD) AND EQUIPMENT IMPROVEMENT REPORT AND MAINTENANCE SUMMARY (EIR MS). The quarterly Equipment Improvement Report and Maintenance Digest, TB 43-0001-39 series, contains valuable field information on the equipment covered in this manual. The information in the TB 43-0001-39 series is compiled from some of the Equipment Improvement Reports that you prepared on the engines covered in this manual. Many of these articles result from comments, suggestions, and improvement recommendations that you submitted to the EIR program. The TB 43-0001-39 series contains information on equipment improvements, minor alterations, proposed Modification Work Orders (MWO's), warranties (if applicable), actions taken on some of your DA Form 2028's (Recommended Changes to Publications), and advance information on proposed changes that may affect this manual. In addition, the more maintenance significant articles, including minor alterations, field-fixes, etc, that have a more permanent and continuing need in the field are republished in the Equipment Improvement Report and Maintenance Summary (EIR MS) for TACOM Equipment (TM 43-0143). Refer to both of these publications (TB 43-0001-39 series and TM 43-0143) periodically, especially the TB 43-0001-39 series, for the most current and authoritative information on your equipment. The information will help you in doing your job better and will help in keeping you advised of the latest changes to this manual. Also refer to DA Pam 310–4, Index of Technical Publications, and Appendix A, References, of this manual.

1-10. REPORTING IMPROVEMENT RECOMMENDATIONS. If your equipment needs improvement, let us know. Send us an EIR. You, the user, are the only one who can tell us what you don't like about your equipment. Let us know why you don't like the design. Tell us why a procedure is hard to perform. Put it on an SF 368 (Quality Deficiency Report) . Mail it to us at: Commander, U.S. Army Tank-Automotive Command, ATTN: DRSTA-MT. Warren, Michigan 48090. We'll send you a reply.

1-11. METRIC SYSTEM. The equipment /system described herein is nonmetric and does not require metric common or special tools. Therefore, metric units are not supplied. Tactical instructions, for sake of clarity, will also remain nonmetric.

1-12. DESTRUCTION TO PREVENT ENEMY USE. Follow procedures given in TM 750-244-6 for destruction of Army material to prevent enemy use.

1-13. ADMINISTRATIVE STORAGE. Requirements for Administrative Storage are covered in TM 740-90-1.

Table 1-1. Standard Torque Specifications

USAGE	MUCH USED	MUCH USED	USED AT TIMES	USED AT TIMES
CAPSCREW DIAMETER AND MINIMUM TENSILE STRENGTH PSI [KG/SQ CM]	To 1/2–69,000 [4850.7000] To 3/4–64,000 [4499.2000] To 1 –55,000 [3866.5000]	To 3/4–120,000 [8436.0000] To 1 –115,000 [8084.5000]	To 5/8–140,000 [9842.0000] To 3/4–133,000 [9349.9000]	150,000 [10545.0000]
QUALITY OF MATERIAL	INDETERMINATE	MINIMUM COMMERCIAL	MEDIUM COMMERCIAL	BEST COMMERCIAL
SAE GRADE NUMBER	1 or 2	5	6 or 7	8

CAPSCREW HEAD MARKINGS

Manufacturer's marks may vary
These are all SAE Grade 5
(3 line)

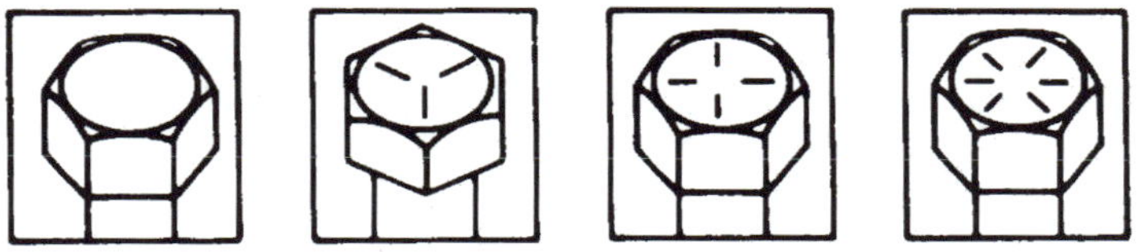

CAPSCREW BODY SIZE (INCHES)–(THREAD)	TORQUE FT-LB [KG M]	TORQUE FT-LB [KG M]	TORQUE FT-LB [KG M]	TORQUE FT-LB [KG M]
1/4–20	5 [0.6915]	8 [1.1064]	10 [1.3830]	12 [1.6596]
–28	6 [0.8298]	10 [1.3830]		14 [1.9362]
5/16–18	11 [1.5213]	17 [2.3511]	19 [2.6277]	24 [3.3192]
–24	13 [1.7979]	19 [2.6277]		27 [3.7341]
3/8–16	18 [2.4894]	31 [4.2873]	34 [4.7022]	44 [6.0852]
–24	20 [2.7660]	35 [4.8405]		49 [6.7767]
7/16–14	28 [3.8132]	49 [6.7767]	55 [7.6065]	70 [9.6810]
–20	30 [4.1490]	55 [7.6065]		78 [10.7874]
1/2–13	39 [5.3937]	75 [10.3725]	85 [11.7555]	105 [14.5215]
–20	41 [5.6703]	85 [11.7555]		120 [16.5960]
9/16–12	51 [7.0533]	110 [15.2130]	120 [16.5960]	155 [21.4365]
–18	55 [7.6065]	120 [16.5960]		170 [23.5110]
5/8–11	83 [11.4789]	150 [20.7450]	167 [23.0961]	210 [29.0430]
–18	95 [13.1385]	170 [23.5110]		240 [33.1920]
3/4–10	105 [14.5215]	270 [37.3410]	280 [38.7240]	375 [51.8625]
–16	115 [15.9045]	295 [40.7985]		420 [58.0860]
7/8–9	160 [22.1280]	395 [54.6285]	440 [60.8520]	605 [83.6715]
–14	175 [24.2025]	435 [60.1605]		675 [93.3525]
1–8	235 [32.5005]	590 [81.5970]	660 [91.2780]	910 [125.8530]
–14	250 [34.5750]	660 [91.2780]		990 [136.9170]

1. Always use the torque values listed above when specific specifications are not available.

Note: Do not use above values in place of those specified in the engine groups of this manual; special attention should be observed in case of SAE Grade 6, 7 and 8 capscrews.

2. The above is based on use of clean and dry threads.

3. Reduce torque by 10% when engine oil is used as a lubricant.

4. Reduce torque by 20% if new plated capscrews are used.

Caution: Capscrews threaded into aluminum may require reductions in torque of 30% or more, unless inserts are used.

TA 113439

Table 1-2. Special Tools and Equipment

Item	Part No.	National Stock No.	Reference Paragraph	Use
BRUSH, CLEANING	ST 876	7920-00-168-3244	4-6	Used to clean oil passages in crankcase.
COMPRESSOR, PISTON RING	C10899159	5120-00-870-6917	5-9	Used to compress piston rings.
PLUG	10899178	5120-00-870-6914	3-43	Used with puller 5120-00-338-6721 to take off damper and pulley assembly on all engines except LDS-465-2.
PULLER KIT	8708724	5120-00-338-6721	3-43	Used with plug 5120-00-870-6914 to take off damper and pulley assembly on all engines except LDS-465-2.
			3-46	Used to take off camshaft and crankshaft gear.
PULLER, MECHANICAL: (Crankshaft Damper and Pulley)	11642009	5120-00-134-7463	3-43	Used to take off damper and pulley assembly on engine LDS-465-2.
REAMER, HAND	12254220	5110-01-050-2240	4-19	Used to ream intake and exhaust valve guides.
TOOL, REPLACING: (Valve Seat)	11642006	5120-00-134-7480	4-19	Used to put exhaust valve seats into cylinder head.
TOOL, REPLACING: (Valve Seat)	11642007	5120-00-134-7473	4-19	Used to put intake valve seats into cylinder head.
WRENCH, BOX: (Cylinder Head Nuts)	D-10951485	5120-00-930-6346	5-11	Used with torque wrench to tighten cylinder head nuts.
WRENCH PULLEY ADJUSTING: (Air Compressor Pulley)	D-10935288	5120-00-070-7809	3-12	Used to adjust air compressor pulley.

1-14. CALIBRATION. The fuel injection pump assembly must be calibrated, on a fuel injection pump test stand, before it is put on the engine. Refer to TM 9-2910-226-34&P for fuel injection pump calibration.

1-15. DESCRIPTION.

a. Type. All models of the engine, shown in figures 1-1 through 1-11, are four stroke cycle, in line, 6-cylinder, overhead valve, liquid-cooled, compression-ignition engines. The LD models are naturally aspirated, and the LDS and LDT models are turbocharged.

b. Fuel Requirements. The engines will operate satisfactorily on all fuels, as given in paragraph 1-19. No changes or adjustments to the engines are needed when changing grades of fuel.

c. Model Differences. To tell the difference between models, paragraphs and illustrations will be titled to show their application to a particular model.

1-16. LOCATION OF ENGINE COMPONENTS. In this manual, the following terms will be used to identify and locate engine parts and assemblies.

a. Engine Ends. The ends of the engine will be called the "fan end" or "front" and the "flywheel end" or "rear."

b. Engine Sides. Looking from the rear end of the engine toward the front, the side to the right is called the "right side." The side to the left is called the "left" side."

c. Cylinder Numbering. The cylinders are referred to as No. 1 through No. 6, starting from the front of the engine.

d. Main Bearing Numbering. The main bearings are numbered 1 through 7, starting from the front of the engine.

e. Connecting Rod Numbering. The connecting rods are numbered from the front, 1 through 6, and are identified by match numbers stamped on the connecting rod and connecting rod cap bosses.

f. Crankshaft and Camshaft Rotation. The crankshaft turns to the right and camshaft turns to the left, looking from the front of the engine.

g. Fuel Filters. The fuel filters, located at the left rear and rear of the engine, have a replaceable element.

h. Oil Filters. The two oil filters are located on the left side of the engine and are the same and can be changed with each other.

i. Cylinder Heads. The front and rear cylinder heads can be changed with each other. The cylinder heads will be referred to as front cylinder head and rear cylinder head.

j. Water Outlet Manifolds. The two cylinder head water outlet manifolds are the same and can be changed with each other. The manifolds will be referred to as the front cylinder head water outlet manifold and the rear cylinder head water outlet manifold.

k. Rocker Arms, Shafts, and Supports. The assembled rocker arms, shafts, and supports can be changed with each other except for the Model LDS-465-2 intake rocker arms which have an oil squirt hole. The Model LDS-465-2 intake rocker arm may be used when replacing intake rocker arms on other LDS models. The rocker arms will be referred to as the front set of rocker arms and the rear set of rocker arms.

1-17. DIFFERENCES BETWEEN MODELS. Table 1-3 gives a listing of the major differences between engine models.

Table 1-3. Differences Between Models

Engine models	LDS-465-2	LDS-465-1A	LD-465-1	LD-465-1C	LDT-465-1C	LDS-465-1
Brake HP rating	195 at 2800 rpm	175 at 2600 rpm	126 at 2600 rpm	130 at 2600 rpm	130 at 2600 rpm	175 at 2600 rpm
Hydraulic pump	Yes	Yes	No	No	No	Yes
Hydraulic pump oil reservoir	Yes	No	No	No	No	No
Turbocharger	Yes	Yes	No	No	Yes	Yes
Clutch	No	Yes	Yes	Yes	Yes	Yes
Starter location	Right side	Left side	Right side	Right side	Right side	Left side
Generator type	Internally rectified to DC 30-80 amperes	DC 25 amperes	DC 25 amperes	Internally rectified to DC 60 amperes	Internally rectified to 60 DC amperes	DC 25 amperes
Generator strap	Straight	Straight	Straight	Curved	Curved	Straight
Accessory drive	6 belt with idler arm	3 belt with adjusting strap	3 belt with adjusting strap	3 belt with adjusting strap	3 belt with adjusting strap	3 belt with adjusting strap
Pressurization system and breather valve	Yes	No	No	No	No	No
Fuel shutoff type	Electric solenoid	Mechanical	Mechanical	Mechanical	Mechanical	Mechanical
Fuel filters on engine	2 separate filters	1 dual filter	1 dual filter	1 dual filter	1 dual filter	1 single filter
Flywheel drive face	Deep ring wit bout drain holes	Medium ring with 12 drain holes	Shallow ring with or without 12 drain holes	Shallow ring with 12 drain holes	Shallow ring with 12 drain holes	Medium ring with 12 drain holes
Transmission adapter	No	No	Yes	Yes	Yes	No
Piston design	With oil cooling channel	With oil cooling channel	Without oil cooling channel	Wit bout oil cooling channel	Without oil cooling channel	Wit h oil cooling channel
Oil pan deep sump location	Rear	Rear	Front	Front	Front	Rear
Oil level gage location	Right rear	Right rear	Right front	Right front	Right front	Right rear

Table 1-3. Differences Between Models - Cont

Engine models	LDS-465-2	LDS-465-1A	LD-465-1	LD-465-1C	LDT-465-1C	LDS-465-1
Fuel injector nozzles	One hole	One hole	Two holes	Two holes	Two holes	Two holes
Fuel injection pump timing static	22°btc	20°btc	25°btc	25°btc	20°btc	20°btc
Flame heater and elbow	Integral with elbow	Integral with elbow	Integral with elbow	Integral with elbow	Integral with elbow	Separate housing
Flame heater fuel pump mounting	Clamped to bracket on intake manifold	Clamped to water outlet manifold	Clamped to water outlet manifold	Clamped to bracket on intake manifold	Clamped to bracket on intake manifold	Bracket mounted on oil pan rail
Flame heater solenoid valves	No	No	No	No	No	Yes
Flow control nozzle and valve assembly	With check valve	With check valve	With check valve	With check valve	With check valve	Without check valve
Flame heater fuel filter	Throwaway	Throwaway	Throw-away	Throwaway	Throwaway	Replaceable
Flame heater ignition unit location	Clamped to bracket on intake manifold	Clamped to water outlet manifold	Clamped to water outlet manifold	Clamped to bracket on intake manifold	Clamped to bracket on intake manifold	Clamped to intake manifold elbow
Air com-pressor discharge outlet	To rear	To front	To side	To side	To aide	To front

1-18. TABULATED DATA.

a. General . Refer to the organizational maintenance manual for the tabulated data having to do with the general information and performance of the engine. The tabulated data covering the engine accessories is given in the component technical manuals. Refer to appendix A for a listing of these manuals.

b. Engine Identification. The engine identification plate is at the right front top of the crankcase below the exhaust manifold. The identification plate, on all models, identifies the engine as multifuel and gives the manufacturer's name, model number, part number, serial number, and contract number. On some models the identification plate has extra information such as valve lash, static injection advance, and fuel and oil specifications. The engine identification plate on Model LDT-465-1C, built to conform to US Environmental Protection Agency regulations, contains extra information having to do with engine exhaust emission controls.

c. Data Application. Table 1-4 and the following data applies to models given below or to all models if not identified.

Table 1-4. Data Application

Models	LDS-465-2	LDS-465-1A	LDS-465-1	LD-465-1C	LDT-465-1C	LDS-465-1
Dimensions: (inches)						
Length	50.39	48.28	48.04	48.04	48.04	48.28
Width	29.85	32.88	29.68	29.68	30.18	35.36
Height	42.93	40.32	40.59	41.20	41.20	40.59
Weight: (lbs)	1633	1650	1614	1614	1350	1561
Speed: (rpm)						
Governed: full load	2800-2850	2600-2650	2600-2650	2600-2650	2600-2650	2600-2650
Governed: no load (max)	3050-3100	2850-2900	2850-2900	2850-2900	2850-2900	2850-2900
Idle (rein)	650-700	650-700	650-700	650-700	Refer to note	650-770
Cranking, 60°F 24V (min)	200	200	200	200	200	200
Horsepower, all fuels: rpm	2800	2600	2600	2600	2600	2600
Gross horse-power: (min)	195 hp	175 hp	126 hp	126 hp	130 hp	175 hp
Torque, all fuels: rpm	2000	2000	1400	1400	15[illegible]0	2000
Gross torque: (min)	425 pound-feet	425 pound-feet	300 pound-feet	300 pound-feet	305 pound-feet	425 pound-feet

NOTE: Refer to injection pump identification plate for correct idle speed setting.

Cylinders:
- Number 6
- Arrangement .... In-line
- Numbering (from front) 1,2,3,4,5,6
- Firing order 1,5,3,6,2,4
- Bore .... 4.56 in.
- Displacement .... 478 cu in.

Induction System:
- Models LDS-465-2, LDS-465-1, LDS-465-1A, LDT-465-1C .... Turbocharged
- Models LD-465-1, LD-465-1C .... Naturally aspirated

Air flow (at rated power and speed):	
Model LDS-465-2	550 cfm
Models LDS-465-1, LDS-465-1A	450 cfm
Models LD-465-1, LD-465-1C	300 cfm
Model LDT-465-1C	405 cfm
Pistons:	
Stroke	4.87 in.
Compression ratio	22:1
Cooling:	
Type	Liquid cooled, forced circulation, pressure type, thermostat, belt driven water pump and fan.
Engine operating temperature:	
Models LDS-465-1, LDS-465-1A, LDS-465-2	160°F. to 180°F. at 60°F. ambient
Models LD-465-1, LD-465-1C, LDT-465-1C	180°F. to 200°F. at 60°F. ambient
Water pump type	Centrifugal
Water pump output at rated engine speed:	
Model LDS-465-2	180°F. at 2,800 rpm, 95 gpm
Models LDS-465-1, LDS-465-1A, LDT-465-1C, LD-465-1, LD-465-1C	180°F. at 2,600 rpm, 85 gpm

NOTE

Early production Model LDT-465-1C fuel injection pumps had an idle speed setting of 650-700 rpm. Late Production LDT-465-1C injection pumps had an idle speed setting of 800-850 rpm. Refer to injection pump identification plate for correct idle speed setting.

Drive	V-belts
Fan	6 blades
Drive (from crankshaft):	
Models LDS-465-1, LD-465-1C, LDT-465-1C, LDS-465-1A	Clutch drive from face of flywheel
Model LDS-465-2	Automatic transmission from flexible plates on flywheel
Crankshaft rotation (viewed from front)	Clockwise

Fuel System:	
Type	Fuel injection, compression ignition, multifuel
Fuel:	
Refer to paragraph 1-19 for authorized fuels.	
Fuel supply pressure at fuel injection pump inlet at rated speed (rein)	60 psi
Fuel Consumption at rated power and speed (all fuels, max):	
Model LDS-465-2	90 lb/hr
Models LDS-465-1, LDS-465-1A	80 lb/hr
Models LD-465-1, LD-465-1C	61 lb/hr
Model LDT-465-1C	64 lb/hr
Fuel injection pump automatic advance	7.5° to 9.5 °
Valve clearance (cold or hot engine):	
Exhaust	0.025 in.
Intake	0.010 in.
Lubrication:	
Oil specifications	
+32° to +115°F	MIL-L-2104 (OE/HDO-30)
-10° to +40°F	MIL-L-2104 (OE/HDO-10)
-65° to 0°F	(OES)
Normal oil temperature (after oil cooler) at full load:	
Models LDS-465-2, LDS-465-1, LDS-465-1A	200°F. to 220°F. at 60°F. ambient
Models LD-465-1, LD-465-1C, LDT-465-1C	190°F. to 210°F. at 60°F. ambient
Maximum oil temperature (oil pan sump):	
Model LDS-465-2	265°F.
Models LDS-465-1, LD-465-1, LD-465-1C, LDT-465-1C, LDS-465-1A	260°F.
Oil Pressure (crankcase main gallery):	
Model LDS-465-2	
Idle	OE/HDO-30 at 200°F., 15 psi min
2,800 rpm	OE/HDO-30 at 200°F., 40 psi min
Models LDS-465-1, LDS-465-1A, LD-465-1, LD-465-1C, LDT-465-1C	
Idle	OE/HDO-30 at 200°F., 10 psi min
2,600 rpm	OE/HDO-30 at 200°F., 40 psi min

Oil pump outlet at rated engine speed:	
Model LDS-465-2	OE/HDO-30 at 180°F., 29 gpm
Models LDS-465-1, LD-465-1, LD-465-1C, LDT-465-1C, LDS-465-1A	OE/HDO-30 at 180°F., 22 gpm
Oil capacity:	
Model LDS-465-2 (filters dry)	20 qt
Model LDS-465-2 (filters full)	16 qt
Models LDS-465-1, LD-465-1, LD-465-1C, LDT-465-1C, LDS-465-1A (filters dry)	22 qt
Models LDS-465-1, LD-465-1, LD-465-1C, LDT-465-1C, LDS-465-1A (filters full)	18 qt
Oil system type	Positive, full pressure lubrication
Oil filters	Full flow (replaceable elements)
Manifold heater (cold weather starting):	
Type	Flame type, spark igniter
Spark igniter gap	0.088 to 0.093 in.
Fuel pump	Electric
Fuel	Same as engine

Drive Ratios and Rotation (From Front)

Component	Rotation	Ratio to Crankshaft
Cam shaft	Counterclockwise	0.50:1
Starter	Counterclockwise	11.5:1
Fuel Injection Pump	Clockwise	1.00:1
Air Compressor	Clockwise	0.86:1
Oil Pump	Clockwise	1.22:1
Tachometer Drive	Counterclockwise	0.50:1
Power Steering Pump	Clockwise	1.37:1
Water Pump and Fan:		
Model LDS-465-2	Clockwise	1.23:1
Models LDS-465-1, LD-465-1, LD-465-1C, LDT-465-1C, LDS-465-1A	Clockwise	1.35:1
Generator:		
Model LDS-465-2	Clockwise	1.74:1
Models LDS-465-1, LD-465-1, LD-465-1C, LDT-465-1C, LDS-465-1A	Clockwise	2.00:1

1-19. AUTHORIZED FUELS.

a. General. Multifuel engines will work right on all Primary and Alternate I fuels. Alternate II fuels may need the addition of diesel fuel for good operation. No engine changes or adjustments are needed when changing grades of fuel.

b. Primary Fuels.

(1) Diesel Fuel VV-F-800, Grades DF-A, DF-1, and DF-2 (NATO Code No. F-54) (Fuel Note 1).

(2) Turbine Fuel MIL-T-5624, Grade JP-5 (NATO Code No. F-44).

c. Alternate I Fuels.

(1) Diesel Fuel, MIL-F-16884, (NATO Code Nos. F-75 or F-76) (Fuel Note 1).

(2) Distillate Fuel, MIL-F-24397, ND (NATO Code No. F-85) (Fuel Note 1).

(3) Commercial Diesel Fuel (ASTM D975), I-D and 2-D (Fuel Note 1).

(4) Turbine Fuel, Aviation, Kerosene-Type (NATO Code No. 34).

(5) Turbine Fuel, MIL-T-38219, Grade JP-7.

(6) Commercial Aviation Turbine Fuel (ASTM D1655), Jet A and Jet A-1.

(7) Aviation Gasoline, MIL-G-5572, AVGAS 80/87 (NATO Code No. F-12).

(8) Commercial Aviation Gasoline (ASTM D910), Grade 80/87.

(9) Commercial Gasoline, Leaded, Low-Lead, or Unleaded, where Research Octane Number (RON) is known to be 89 or below, or Anti-knock Index (Fuel Note 2) is known to be 85 or below.

(10) Any mixture consisting only of Primary and/or Alternate I fuels.

d. Alternate II Fuels. Alternate II fuels may need the addition of diesel fuel for satisfactory operation. Methods for determining if addition of diesel fuel is necessary and the amount required are described in paragraph 1-20.

(1) Turbine Fuel MIL-T-5624, Grade JP-4 (NATO Code No. F-40).

(2) Turbine Fuel, Aviation, Naphtha-Type (ASTM D1655), Jet B.

(3) Gasoline, Unleaded/Low-Leaded, VV-G-001690, Special Grade (91/82).

(4) Combat Gasoline, MIL-G-3056, MOGAS (NATO Code No. F-46).

(5) Gasoline, Automotive (NATO Code No. F-50). (Note 1)

(6) Gasoline, VV-G-76, Regular and Premium Grades. (Note 2)

(7) Gasoline, Unleaded/Low-Leaded, VV-G-001690, Regular and Premium Grades. (Note 2)

(8) Aviation Gasoline, MIL-G-5572, AVGAS 100/300 (NATO Code No. F-18). (Note 1)

(9) Commercial Aviation Gasoline (ASTM D910), Grade 100/130. (Note 1)

NOTE 1 These fuels will require addition of diesel fuel.
NOTE 2 Premium grades of these fuels will require diesel fuel additions.

(10) Commercial Gasoline (ASTM D439), Leaded, Low-Lead, or Unleaded, where Research Octane Number (RON) is known to be above 90 or Antiknock Index (Fuel Note 2) known to be above 86.

e. Emergency Fuels. Long operation on the following emergency fuels may cause early clogging of fuel filters and early fouling of fuel injector nozzles. Some emergency fuels may need the addition of diesel fuel for satisfactory performance.

(1) Burner Fuel Oil, VV-F-815, Grades FO-1 and FO-2 (Fuel Note 1).

FUEL NOTE 1

Diesel fuels and burner fuel oils should not be used at temperatures below the following limits; except for VV-F-800, Grades DF-1 and DF-2, which are supplied to regional and seasonal requirements.

Grade	Temperature
1. DF-1	-10°F., or seasonal limit in accordance with VV-F-800.
2. DF-2	+32°F., or seasonal limit in accordance with VV-F-800.
3. DF-A	No limit
4. 1-D and No. 1	-10°F.
5. 2-D and No. 2	+32°F.
6. NATO Code F-75	+15°F.
7. DF-M	+35°F.
8. ND	+40°F.
9. Gasoline and Turbine Fuels to which Diesel fuel has been added.	-25°F.

FUEL NOTE 2

"Antiknock Index" is Motor Octane Number + Research Octane Number and as of the date of publication of this manual, is the "Octane Number" displayed on retail gasoline pumps in CONUS.

1-20. ALTERNATE 11 FUEL CORRECTION PROCEDURES.

a. This procedure is used for telling if the addition of diesel fuel is needed for good engine operation with unknown or low octane fuels (gasoline and turbine fuel).

b. Run the engine at low idle speed (5 minutes minimum for fuel mixing) until water temperature reaches 120°F. Higher water temperatures are acceptable. However, the misfiring described below may go away as the engine reaches operating temperature.

c. With engine at idle speed and transmission in neutral, attempt to accelerate to rated engine speed. Rated engine speed is 2,600 rpm for all models except LDS-465-2 (M656 Series Trucks), which is 2,800 rpm. Smooth, rapid acceleration to rated speed without misfire shows that fuel is good as is. If this condition is not met, addition of diesel fuel is required (up to 30 percent diesel fuel may be necessary).

NOTE

Misfire at higher than rated engine speed is good while performing the above test.

d. Other reasons for the addition of diesel fuel that may happen during normal operation are:

(1) Not being able to accelerate a warm (120°F. minimum water temperature) engine under light loads.

(2) Popping in the exhaust system when accelerating after going down a long down grade.

(3) Popping in the exhaust system when shifting transmission.

NOTE

Vehicles with two tank fuel systems:

If unknown, or low octane fuel is added to both tanks, the secondary tank must have the same amount of diesel fuel added to the primary tank plus an additional 10 gallons of diesel fuel.

CHAPTER 2

PRELIMINARY PROCEDURES FOR ENGINE MAINTENANCE

Section I. SCOPE

2-1. EQUIPMENT ITEMS COVERED. This chapter gives instructions to take out the engine from the shipping and storage container and get it ready for maintenance.

2-2. EQUIPMENT ITEMS NOT COVERED. All equipment items are covered in this chapter.

Section II. PREPARATION OF ENGINE FOR REPAIR

NOTE

This procedure is the same for all engine models except where noted.

TOOLS: No special tools required

SUPPLIES: 5 1/2-gallon container

PERSONNEL: Two

EQUIPMENT CONDITION: Engine in shipping and storage container.

2-3. REMOVAL OF ENGINE FROM SHIPPING AND STORAGE CONTAINER.

NOTE

This task is the same for all engines.
Engine LDS-465-1A is shown .

FRAME 1

1. Unscrew and take out pneumatic tank valve cap (1) .
2. If valve cap (1) has slot , use slotted end of valve cap to unscrew and take out valve core (2) to vent shipping and storage container (3). If valve ca p does not have slot, use valve stem repair tool to unscrew and take out valve core to vent shipping and storage cent airier.
3. Using slotted end of valve cap (1) or valve stem repair tool, screw valve core (2) back into pneumatic tank valve (4) and tighten. Screw valve cap back on finger tight .

GO TO FRAME 2

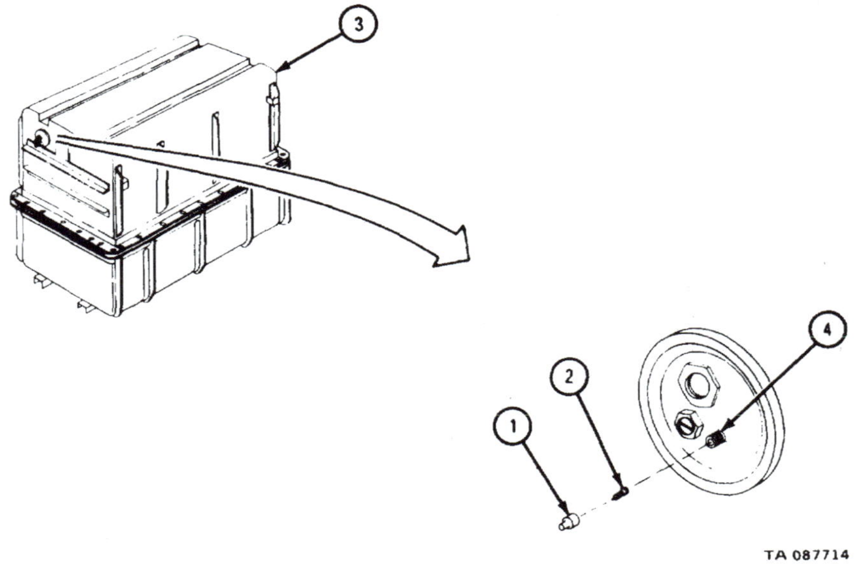

FRAME 2

	1. Takeout 38 screws (1), nuts (2), and lockwashers (3).
Soldier A	2. Hook lifting sling (4) to four shipping container lifting brackets (5), and hook chain hoist (6) to lifting sling as shown.
Soldier B	3. Work chain hoist to lift shipping container upper section (7) from shipping container lower section (8).
	4. Work chain hoist to set shipping container upper section (7) aside.

CAUTION

Be careful not to damage rubber seal (9). It will be used again when engine is put back in shipping container after engine overhaul.

GO TO FRAME 3

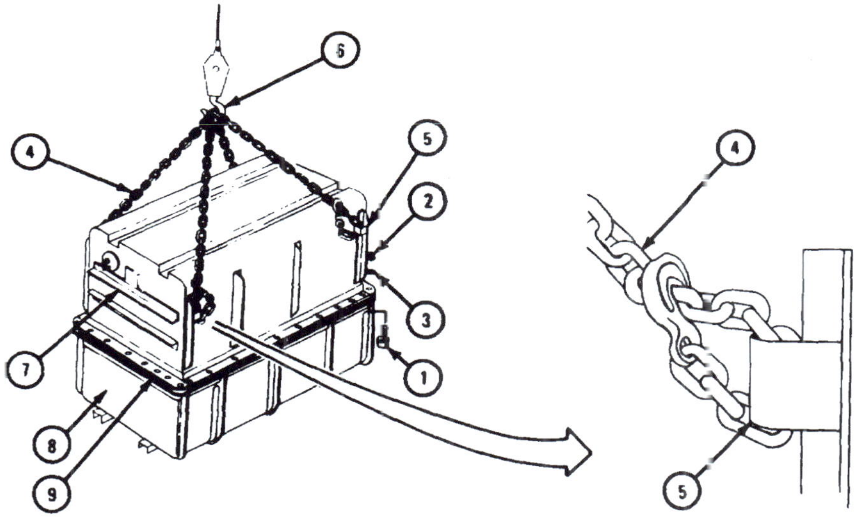

TA 087715

FRAME 3

WARNING

Engine weight is 1600 pounds. Do not use a chain hoist and lifting sling of less than two-ton capacity. Engine could fall and cause injury to personnel.

	1.	Take box (1) holding clutch parts off rear lifting bracket (2) .
Soldier A	2.	Hook chain hoist (3) and lifting sling (4) to front and rear lifting brackets (5 and 2) as shown.
Soldier B	3,	Work chain hoist so lifting sling chain (4) is pulled tight, but do not lift engine and shipping container (6) off ground.

GO TO FRAME 4

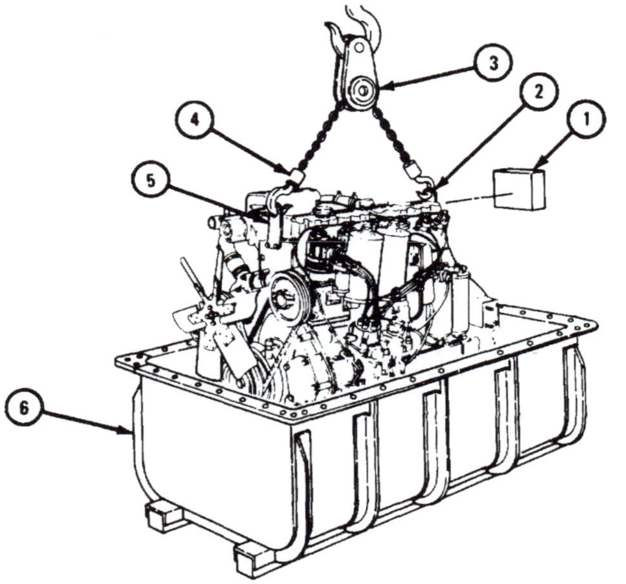

FRAME 4

WARNING

Make sure lifting sling (1) is pulled tight before taking mounting hard ware out of engine mounting brackets in shipping container (2). Engine could slip off suspension frame (3) and fall, causing injury to personnel.

Soldier A 1. Take out two nuts, lockwashers, and screws (4).

2. Take out two nuts, lockwashers, and screws (5).

GO TO FRAME 5

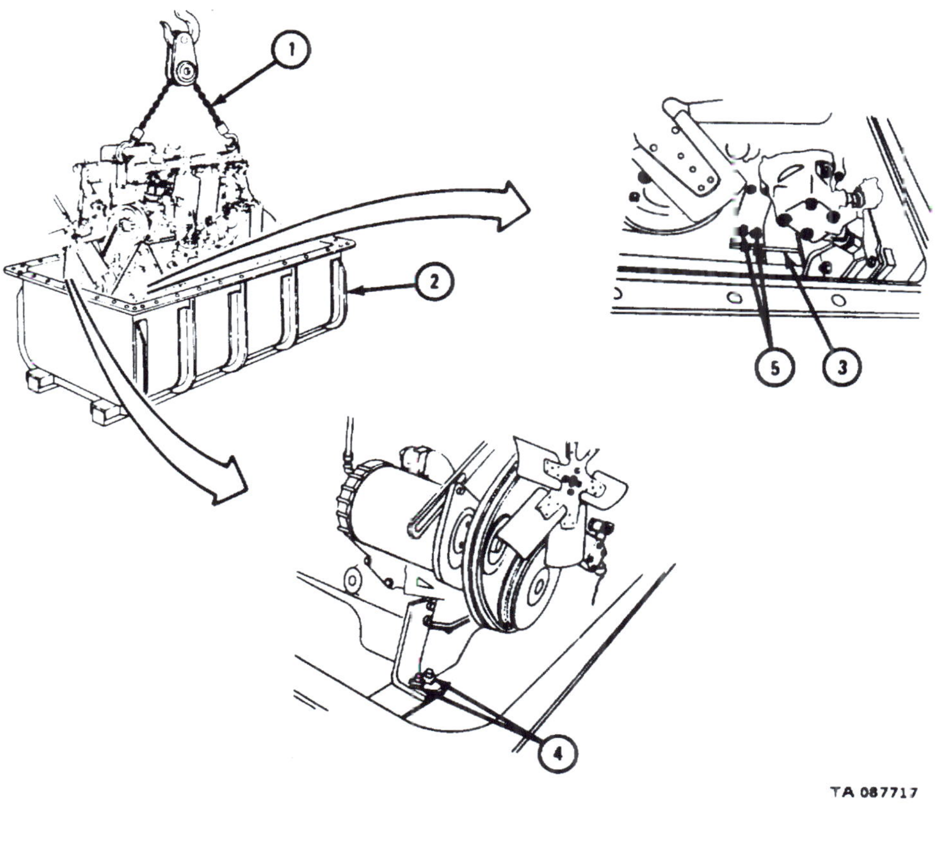

FRAME 5

Soldier A 1. Takeout four screws, lockwashers, and nuts (1) .

2. Guide engine (2) out of shipping container (3) so it does not turn on chain hoist while soldier B works chain hoist to lift out engine.

Soldier B 3. Work chain hoist (4) to lift engine (2) out of shipping container (3) .

GO TO FRAME 6

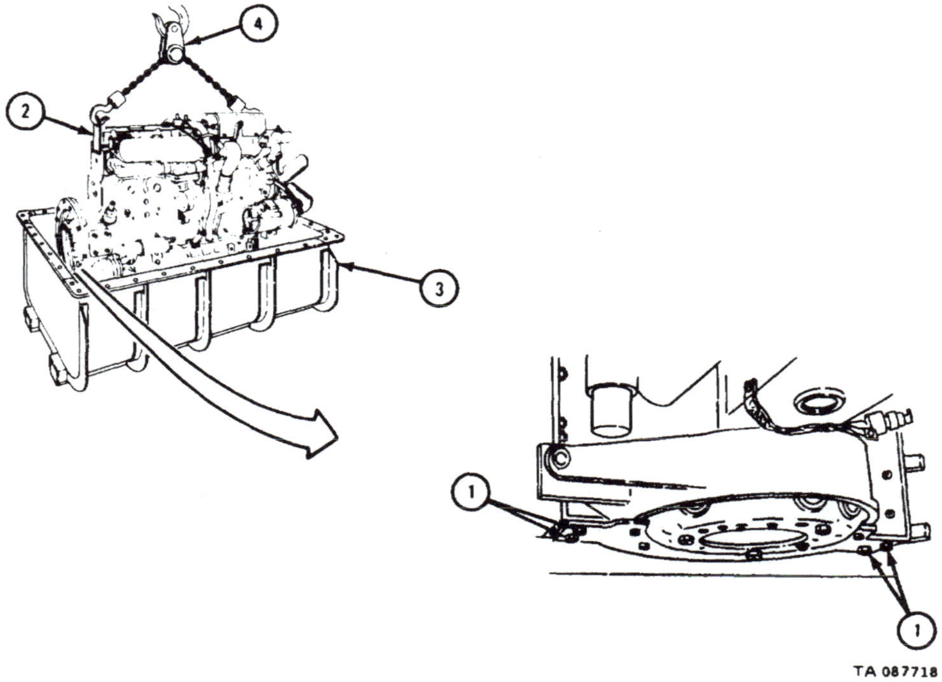

FRAME 6

Soldier A 1. Hold engine (1) so it does not turn on sling (2) while soldier B takes off fan end brackets (3).

Soldier B 2. Take out four screws, lockwashers, and nuts (4) and take off two fan end brackets (3). Pack fan end brackets (3) and screws lockwashers, and nuts (4) in shipping container so they are not lost.

GO TO FRAME 7

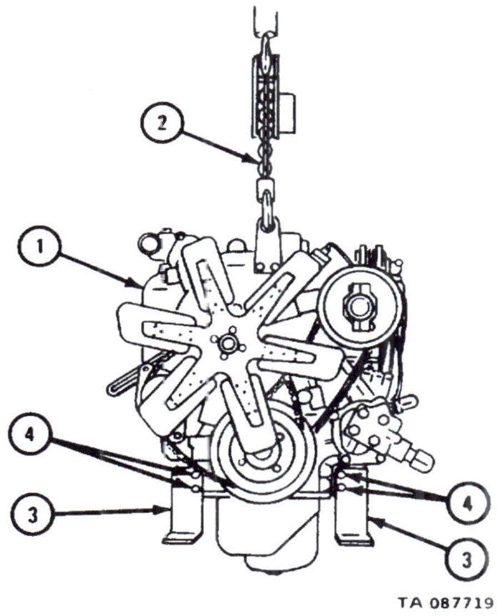

FRAME 7

Soldier A 1. Hold engine (1) so it does not turn on sling (2) when soldier B takes off flywheel endplate (3) .

NOTE

Engine LDS-465-2 has six screws and lockwashers (4) . Engines LDS-465-1 and LDS-465-2A have seven screws and lockwashers (4) . Engines LD-465-1, LD-465-1C , and LDT-465-1C have eight screws and lockwashers (4) .

Soldier B 2. Take out screws and lockwashers (4) and take off flywheel endplate (3) . Pack flywheel endplate and screws and lockwashers in shipping cent airier so they are not lost.

END OF TASK

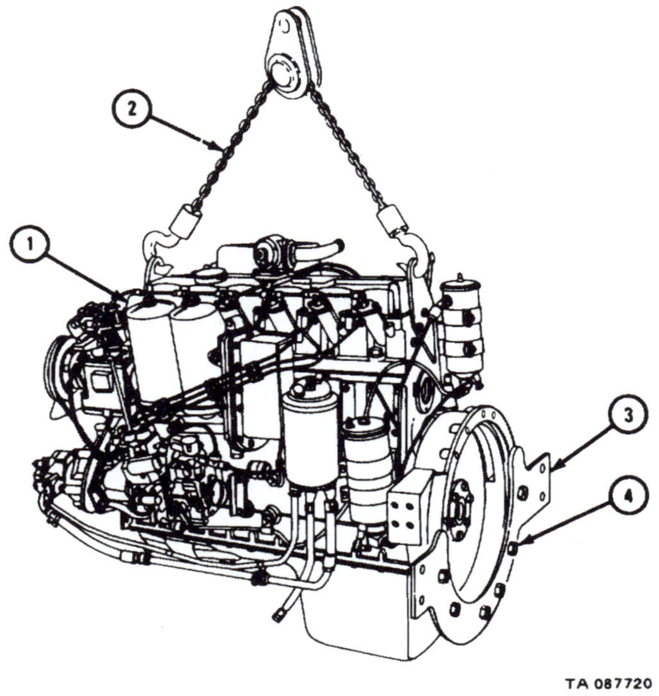

2-4. DRAINING OF ENGINE COOLANT, OIL, AND FUEL.

NOTE

Engine should be drained while it is hanging from lifting sling, before mounting on maintenance stand .

a. Draining Engine Coolant.

FRAME 1

NOTE

Engine may have already been drained before packing in shipping and storage container.

1. Place container below coolant drain cock (1). Open drain cock and let coolant drain out. Close drain cock when coolant stops draining.
2. Put coolant in approved disposal area.

END OF TASK

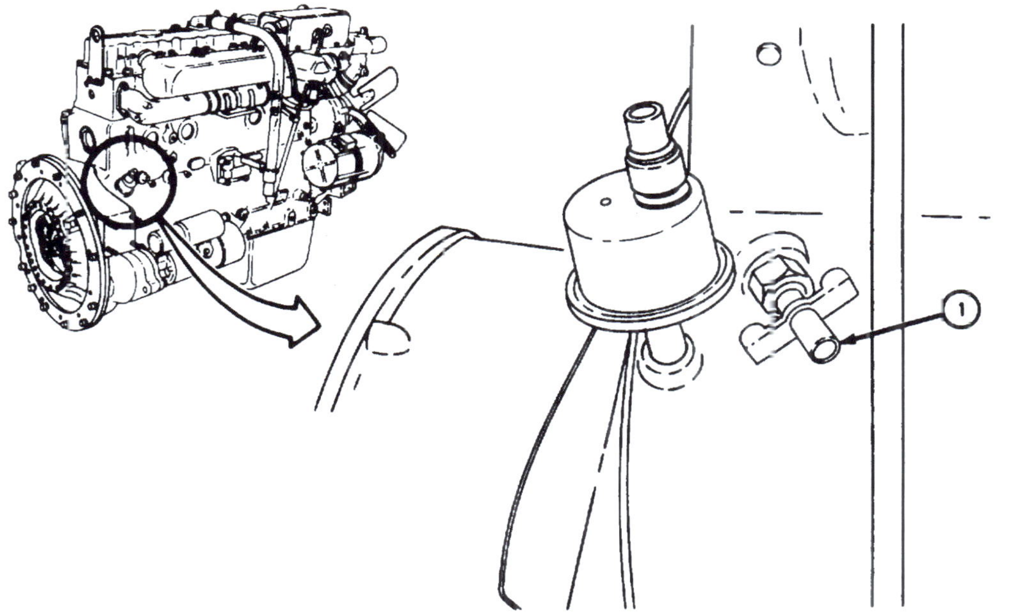

b. Draining Engine Oil.

FRAME 1

NOTE

Engine holds up to 22 quarts (5 1/2 gallons) of oil. Deep oil sump (3) and shallow oil sump (4) are in opposite positions on some engines. Either oil sump can be drained first .

1. Put container under drain plug (1). Take out drain plu g and drain oil into cent airier. Put drain plug back after oil stops draining.
2. Put container under drain plug (2). Take out drain plu g and drain oil into container. Put drain plug back after oil stops draining.
3. Empty container in approved disposal area.

END OF TASK

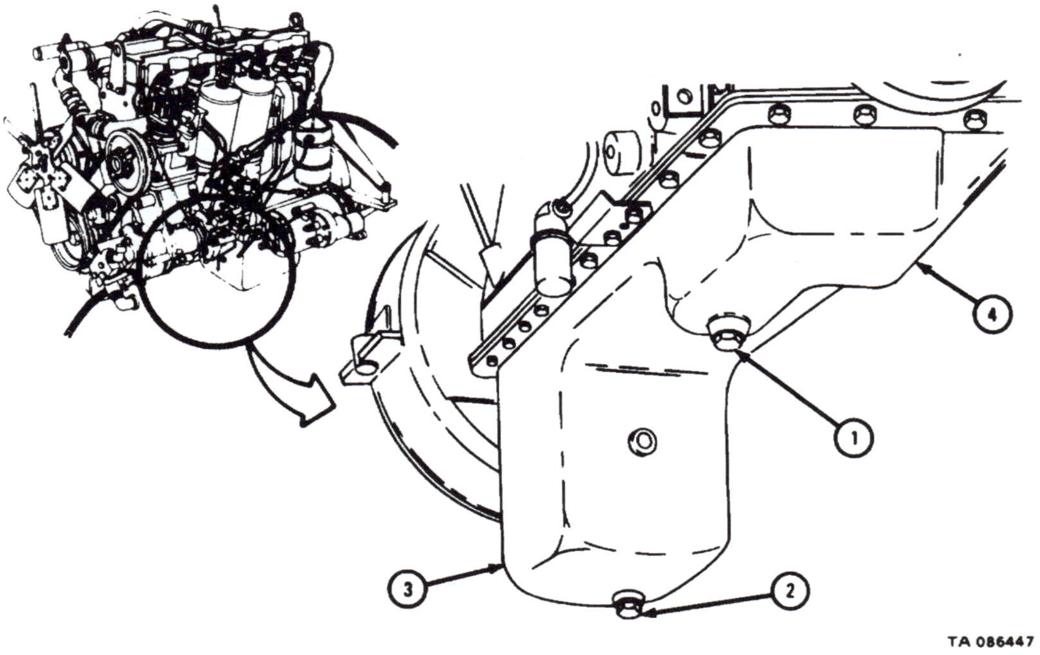

WARNING

Smoking, sparks or open flames are not allowed within 50 feet of work area during this task. Fire or explosion could occur, causing injury to personnel and damage to equipment.

c. Draining Fuel.

(1) Engines LD-465-1, LD-465-1C, LDT-465-1C, and LDS-465-1A.

FRAME 1

NOTE

Fuel filters shown are for engine LDS-465-1A. Procedure is the same for engines LD-465-1, LD-465-1C, and LDT-465-1C.

1. Turn fuel filter vent valve (1) to the left.
2. Place container under two fuel drain cocks (2). Open two drain cocks and drain fuel into container. Close drain cocks and vent valve (1) when fuel Stops draining out.
3. Empty container in approved disposal area.

END OF TASK

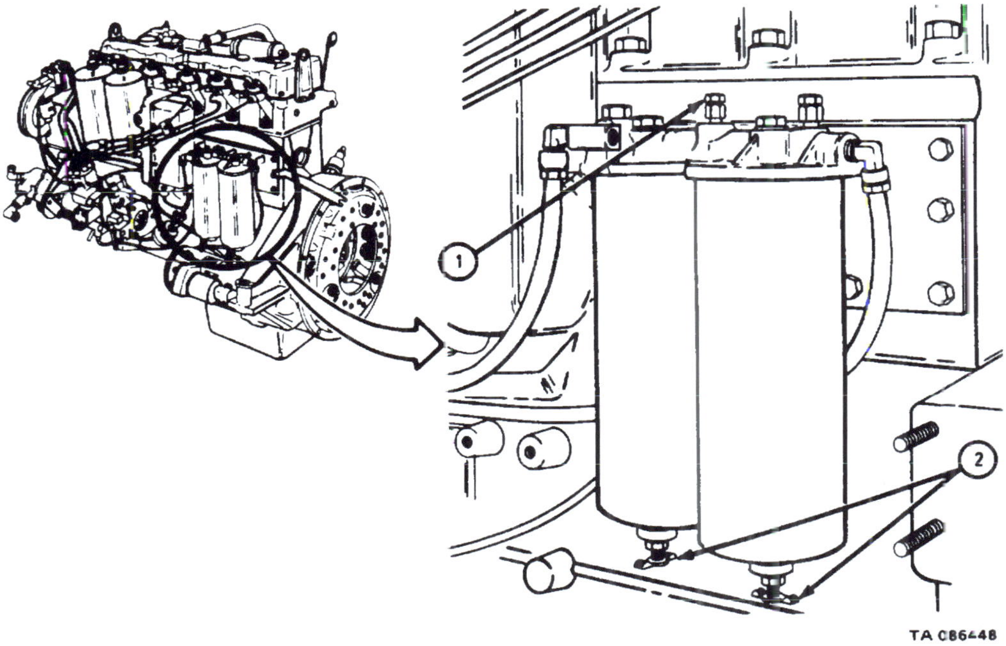

(2) Engine LDS-465-1 .

FRAME 1

1. Turn fuel filter vent valve (1) to the left.
2. Put container under fuel drain cock (2). Open drain cock and drain fuel into container . Close drain and vent valve (1) when fuel stops draining.
3. Empty container in approved disposal area.

END OF TASK

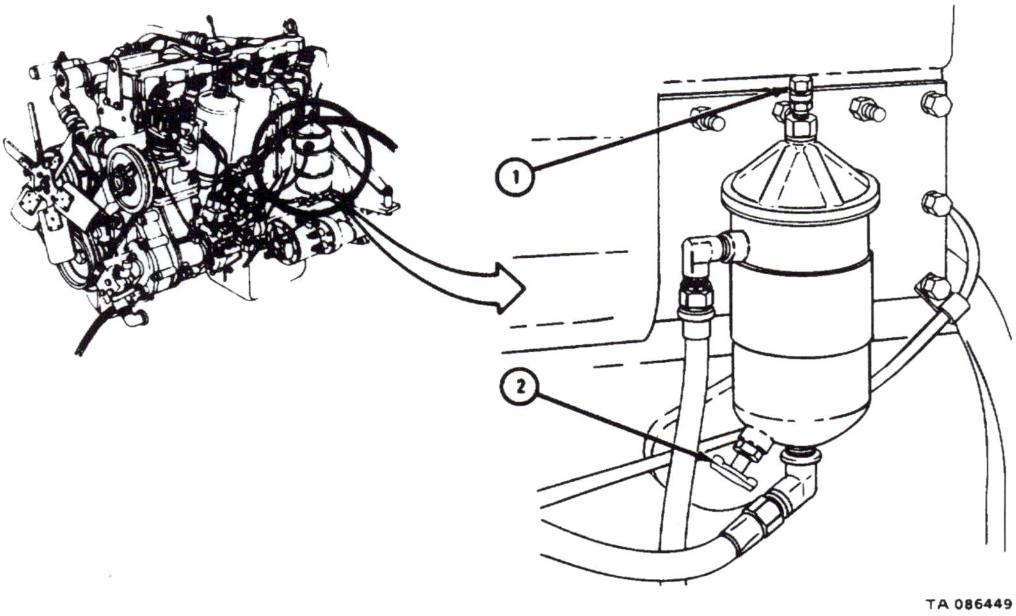

(3) Engine LDS-465-2 .

FRAME 1

1. Turn fuel vent valve (1) on final fuel filter (2) to the left.
2. Place container under end of drain hose (3). Open drain cock (4) and drain fuel into container. Close drain cock when fuel stops draining.
3. Turn fuel vent valve (4) on secondary fuel filter (5) to the left.
4. Place container under fuel drain cock (6). Open drain cock and drain fuel into container . Close drain cock and vent valve (1) when fuel stops draining.
5. Empty container in approved disposal area.

END OF TASK

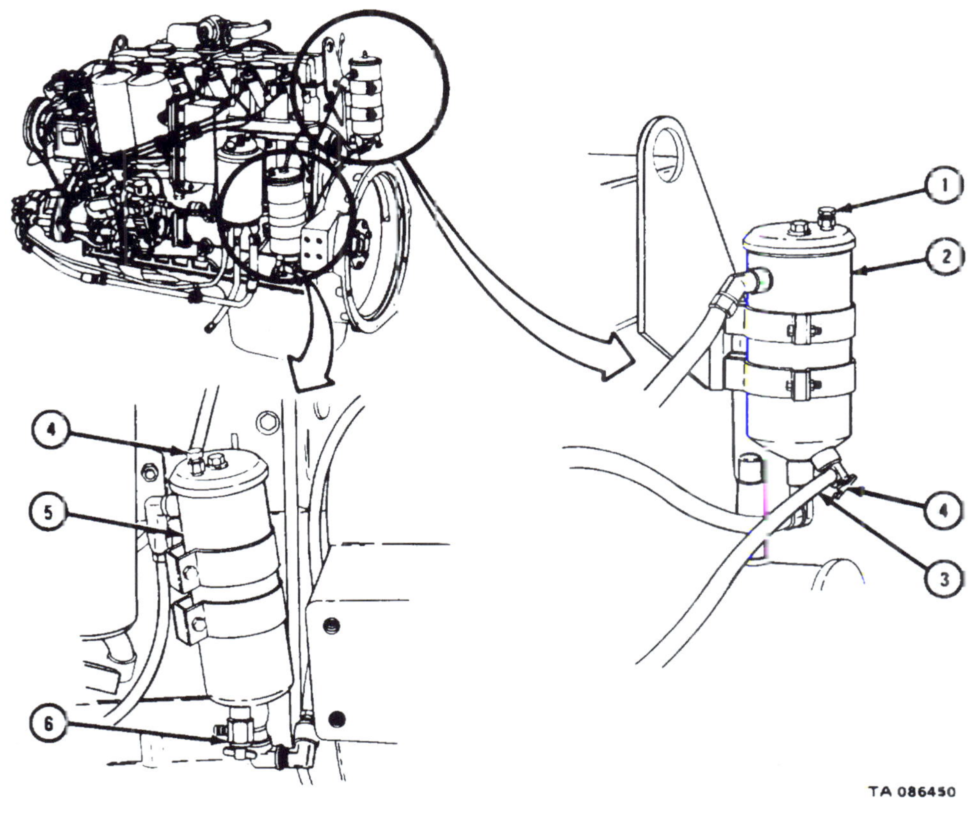

2-5. MOUNTING ENGINE ON TRANSPORT STAND.

FRAME 1

1. Loosen four screws (1) . Slide four engine supports (2) out away from center of engine stand (3) .
2. Position engine stand (3) so engine (4) rests between four engine stand supports (2) .

GO TO FRAME 2

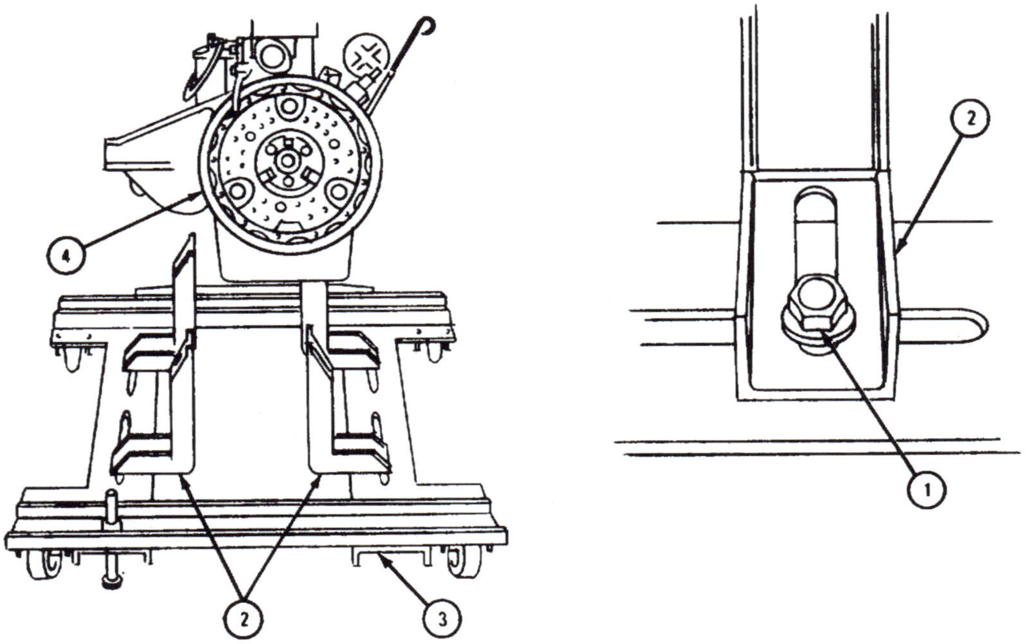

TA 117784

FRAME 2

	1.	Slide four engine supports (1) in place.
Soldier A	2.	As engine (2) is lowered onto stand (3), guide engine so four engine supports (1) will set on oil pan lip (4) as shown.
Soldier B	3.	Lower hoist slowly until oil pan lip (4) just touches four engine supports (1).
Soldier A	4.	Tighten four capscrews (5).
Soldier B	5.	Lower hoist and take off lifting sling.

END OF TASK

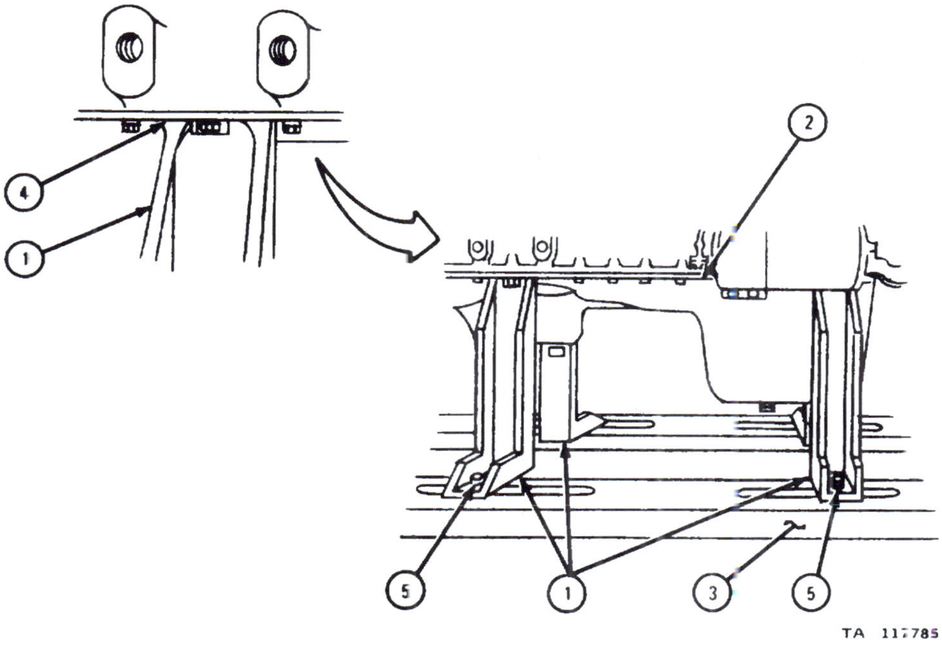

CHAPTER 3

ENGINE DISASSEMBLY

Section I. SCOPE

3-1. EQUIPMENT ITEMS COVERED. This chapter gives instructions for taking apart the engine assembly.

3-2. EQUIPMENT ITEMS NOT COVERED. All equipment items are covered in this chapter.

Section II. REMOVAL OF ENGINE ACCESSORIES

NOTE

This procedure is the same for all engines except where noted.

TOOLS: Air compressor pulley adjusting wrench, pn 10935288

SUPPLIES: Tags
Container
Cap and plug set
Cloth-backed adhesive tape, MIL-T-43566
Hexhead machine screw, 3/8 x 1 1/2-inches
Capscrew, 3/8 x 3/4 inch (2)

PERSONNEL: Two

EQUIPMENT CONDITION: Engine mounted on transport stand.

3-3. STARTER ASSEMBLY.

a. Engines LD-465-1, LD-465-1C, and LDT-465-1C.

FRAME 1

Soldier A 1. Take off three nuts (1) and lockwashers (2).

WARNING

Starter assembly (3) weighs fifty pounds. Be careful to hold it up firmly when taking it off mounting studs. Starter could fall and cause injury to personnel and damage to equipment.

Soldiers A and B 2. Hold starter (3) up and pull it straight out from engine until starter drive gear is free.

Soldier A 3. Take off starter adapter (4) and adapter gasket (5). Throw away starter adapter gasket. Take off and throw away starter gasket (6).

END OF TASK

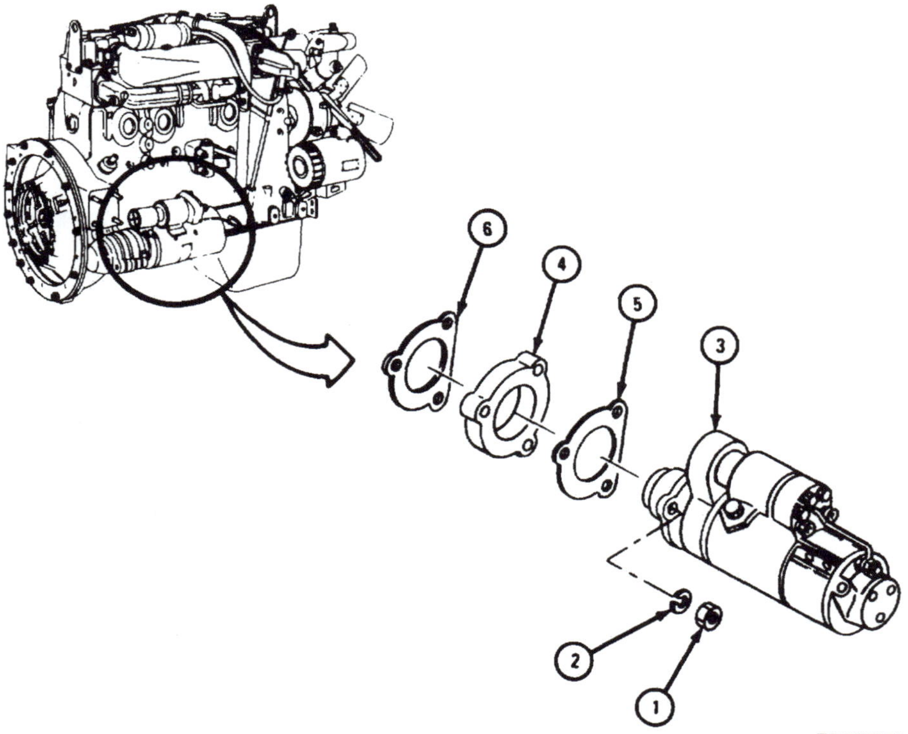

b. Engines LDS-465-1, LDS-465-1A, and LDS-465-2.

FRAME 1

Soldier A	1.	Take off three nuts (1) and lockwashers (2).

WARNING

Starter (3) weighs fifty pounds. Be careful to hold it up firmly when taking it off mounting studs. Starter could fall and cause injury to personnel and damage to equipment.

Soldiers A and B	2.	Hold starter (3) up and pull it straight out from engine until starter drive gear is free.
Soldier A	3.	Take off and throw away starter gasket (4).

END OF TASK

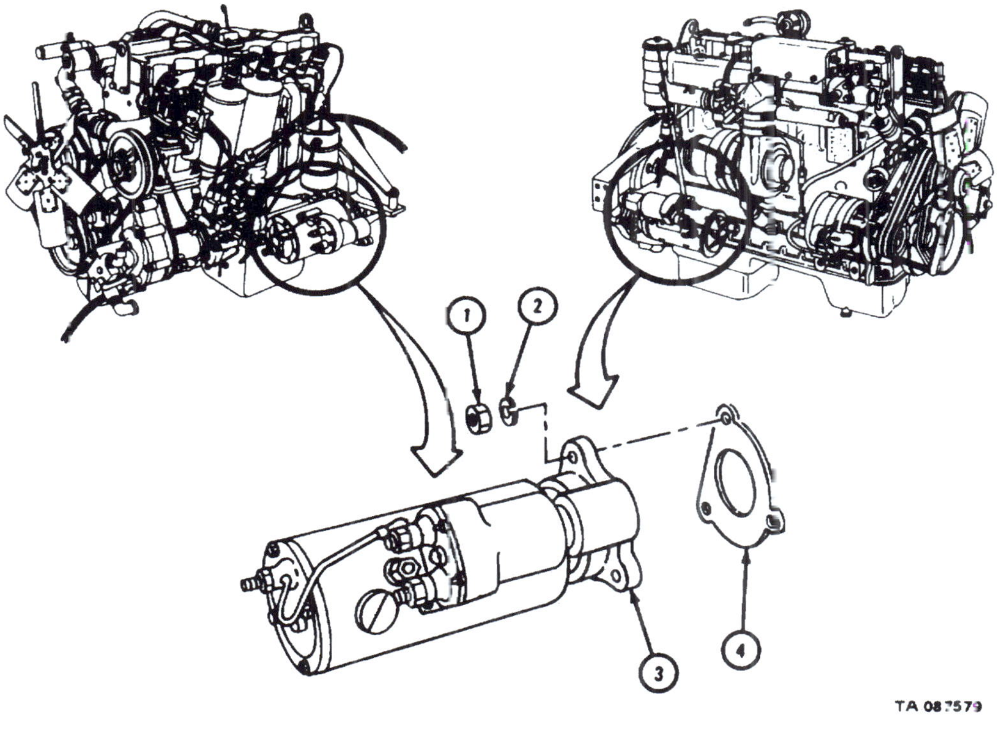

3-4. GENERATOR ASSEMBLY.

a. Engines LD-465-1C and LDT-465-1C.

FRAME 1

1. Take screw, lockwasher, and stepped washer (1) out of adjustment bracket (2).
2. Swing adjustment bracket (2) up and away from generator (3).

CAUTION

Hold up generator (3) when drive belts (4) are taken off so that it does not swing down and hit mounting bracket. Mounting bracket could be damaged.

3. Push generator (3) toward engine and take two drive belts (4) off pulley (5). Tag drive belts.

GO TO FRAME 2

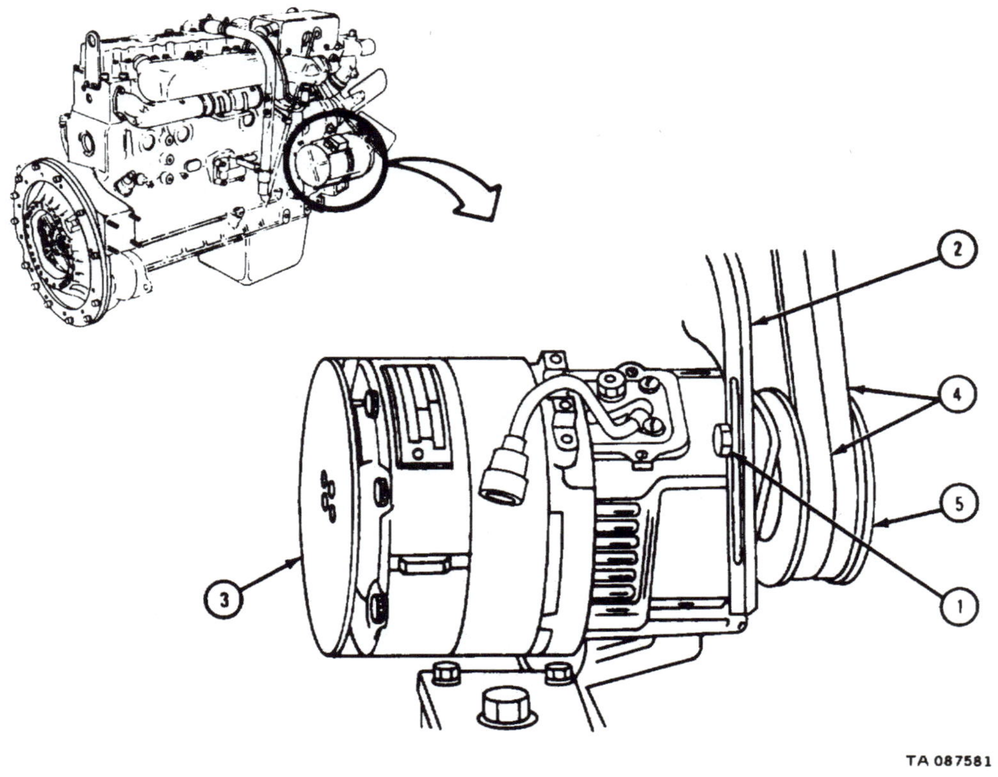

FRAME 2

NOTE

Some engines do not have two cotter pins (1).

1. Take two cotter pins (1) out of two bolts (2) and throw cotter pins away.
2. Take off two nuts and washers (3).

NOTE

Some engines have one or more shims (4) between generator (5) and mounting bracket (6) .

3. If generator (5) has shims (4), put container under mounting bracket (6) to catch shims so they are not lost.
4. Hold generator (5) up and pull out two bolts (2). Leave flat washers (7) on bolts. Lift out generator.
5. Tag shims (4) .

END OF TASK

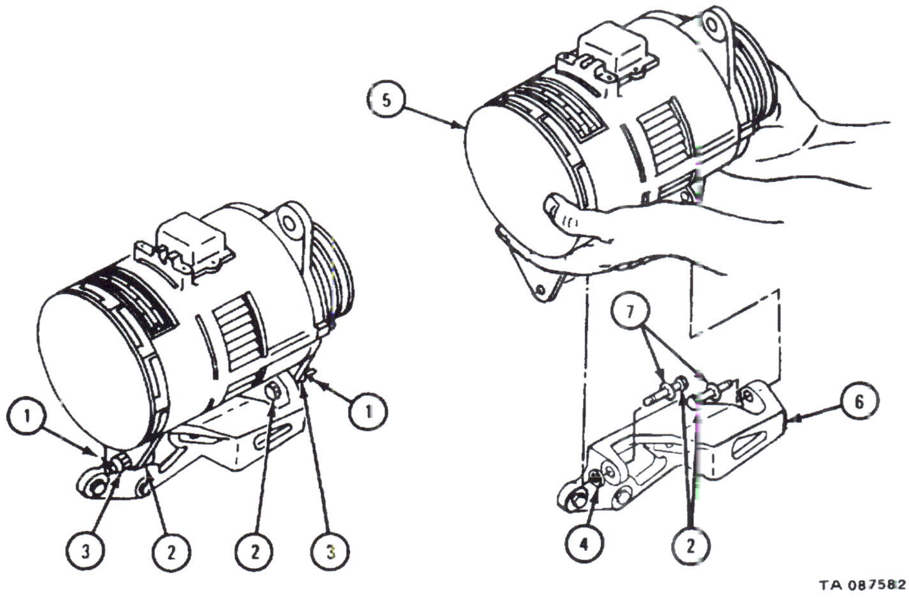

b. Engines LD-465-1, LDS-465-1, and LDS-465-1A.

FRAME 1

1. Take out screw (1), lockwasher (2), and stepped washer (3).
2. Pull bracket (4) up and away from generator (5).

NOTE

Some engines do not have ground strap (6).

3. Take out fillister-head screw and washer (7) and take off ground strap (6).

NOTE

Some engines do not have two cotter pins (8).

4. Pull out and throw away two cotter pins (8).
5. Loosen but do not take out two screws (9).

CAUTION

Hold up generator (5) when drive belts (10) are taken off so that it does not swing down and hit mounting bracket. Mounting bracket could be damaged.

6. Push generator (5) toward engine. Take off and tag two generator drive belts (10).

GO TO FRAME 2

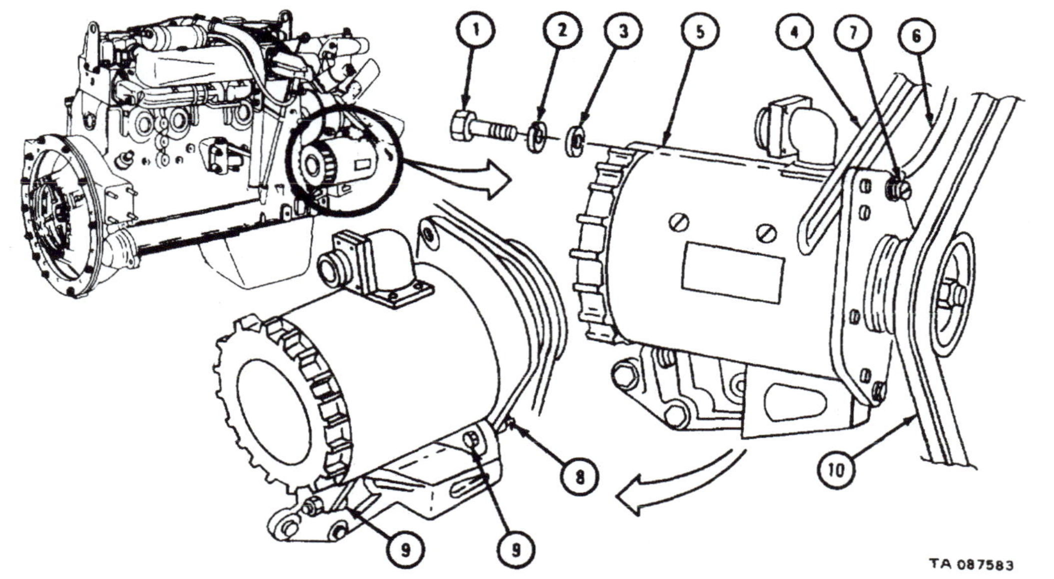

FRAME 2

1. Take off two nuts (1) and washers (2).

NOTE

Some engines have one or more shims (3) between generator (4) and mounting bracket (5) .

2. If generator (4) has shims (3), put container under mounting bracket (5) to catch shims so they are not lost.
3. Hold generator (4) firmly and pull mounting screws (6) out of bracket (5). Leave two washers (7) on screws (6). Lift out generator .
4. Tag shims (3) .

END OF TASK

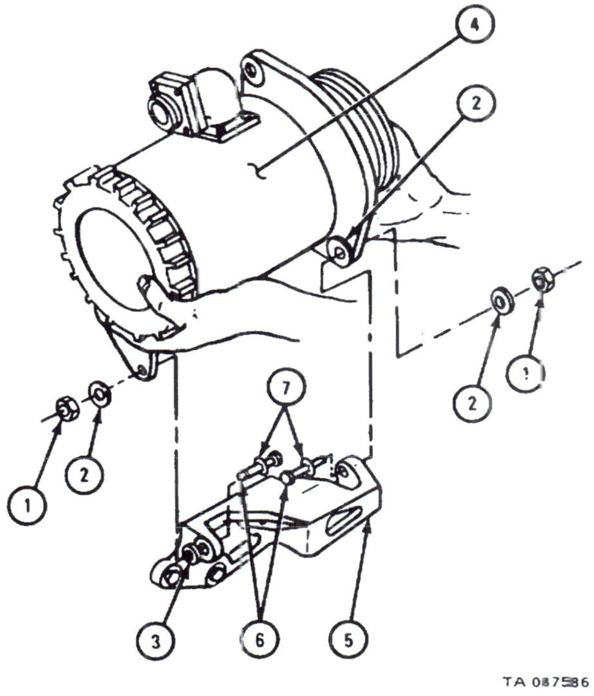

c. Engine LDS-465-2 .

FRAME 7

1. Hold nut (1), unscrew nut (2), and takeoff air pressure hose (3) .
2. Take out screw (4), lockwasher (5), serrated washer (6) .

CAUTION

Hold up generator (7) when drive belts (9) are taken off so it does not swing down and hit side of engine. Damage to generator housing could result.

3. Hold up generator (7) . Loosen two locknuts (8) and swing generator toward engine . Take off and tag three generator drive belts (9) from generator pulley (10) .

GO TO FRAME 2

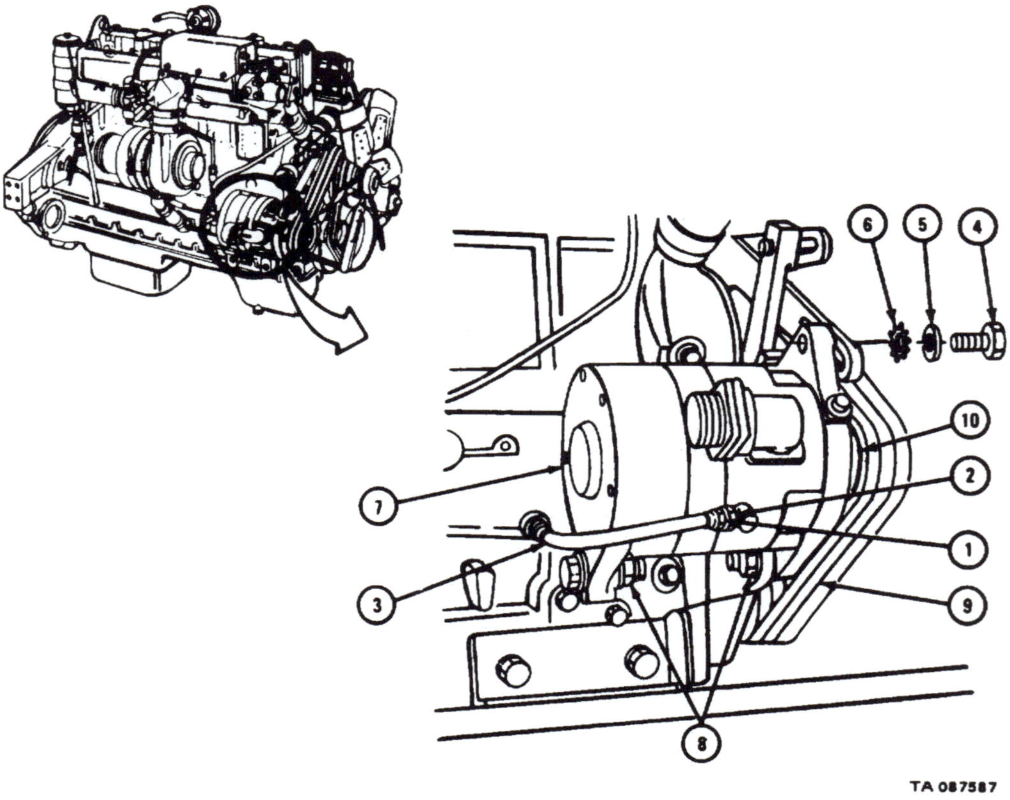

FRAME 2

Soldier A	1.	Hold up generator (1) so it does not fall when two mounting screws (2) are taken out.
Soldier B	2.	Take out two locknuts (3), four washers (4), and two screws (2).
Soldiers A and B	3.	Lift out generator (1)

END OF TASK

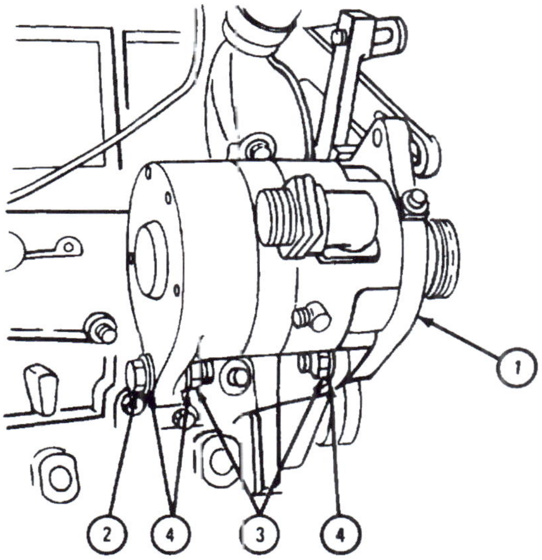

TA 087588

3-5. TURBOCHARGER ASSEMBLY.

a. Engine LDT-465-1C.

FRAME 1

1. Loosen two hose clamps (1) and slide them to center of oil drain hose (2).
2. Take out two screws and lockwashers (3). Take off oil drain tube (4) and oil drain hose (2). Tag oil drain hose.
3. Take off and throw away turbocharger oil outlet gasket (5).
4. Unscrew and take out oil drain hose adapter (6).

GO TO FRAME 2

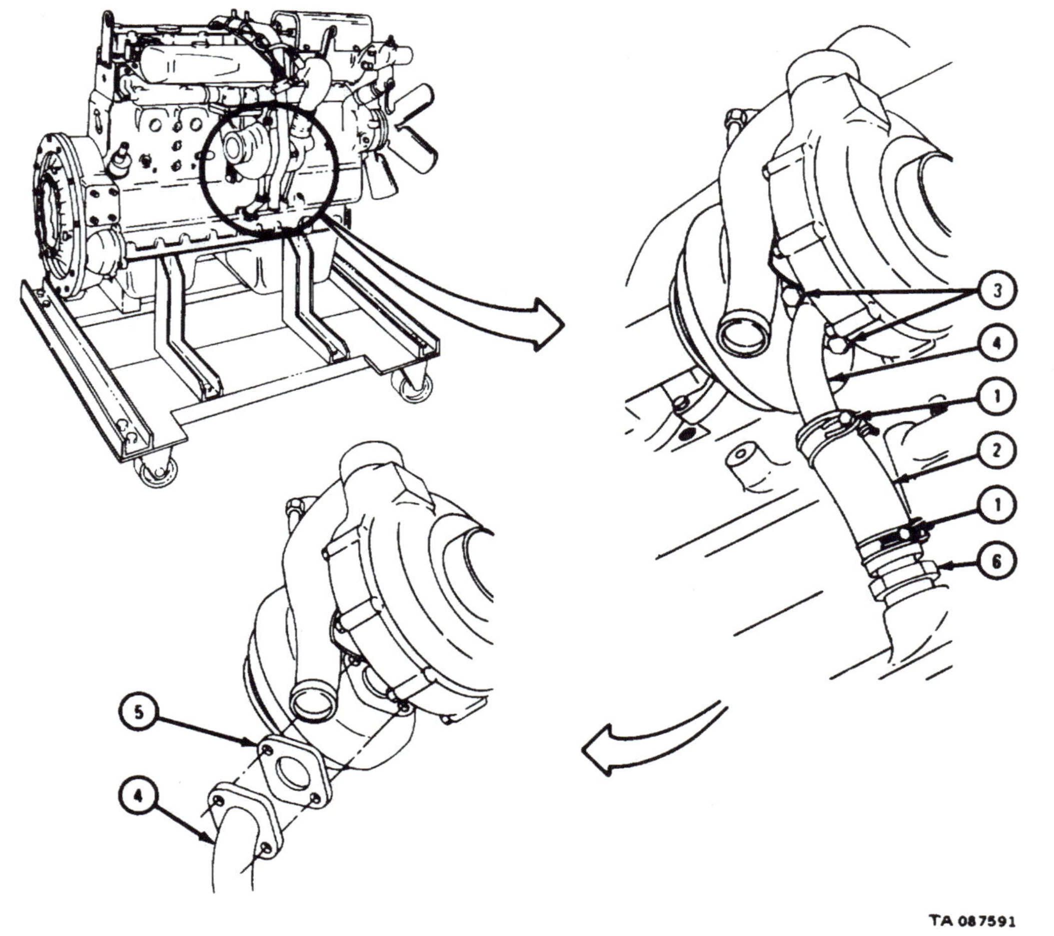

FRAME 2

1. Unscrew ignition lead connector (1). Take off and tag ignition lead (2).
2. Unscrew two uppercuts (3). Take off and tag flame heater fuel inlet tube (4) and flame heater fuel return tube (5).
3. Loosen hose clamp (6).

NOTE

Some engines do not have flame heater cover (7).

4. Take out four screws and starwashers (8) and take off flame heater cover (7).
5. Loosen two screws (9) .
6. Take off four nuts and lockwashers (10), lift up bracket (11) and take off flame heater intake manifold elbow (12).

GO TO FRAME 3

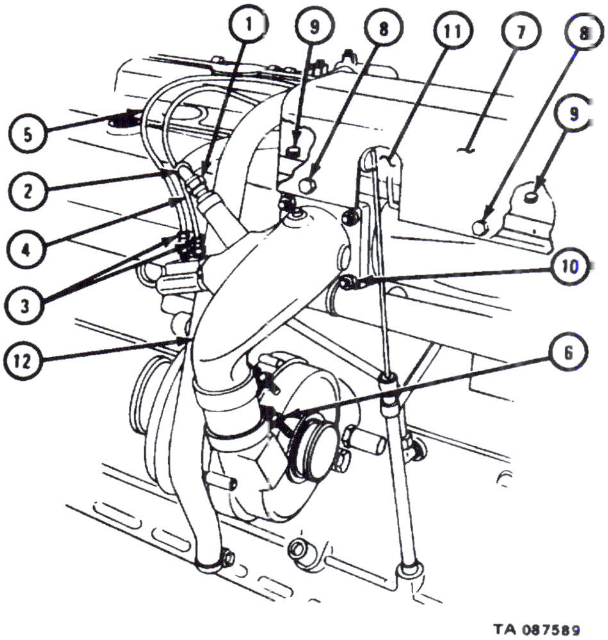

FRAME 3

1. Take off and throw away flame heater intake manifold elbow gasket (l).
2. Loosen hose clamp (2) and take out crankcase breather tube (3).

NOTE

If engine has flame heater cover (4), this cover will be used later to tell what type of flame heater is used on this engine. It must be put back now for htis purpose.

3. Put back flame heater cover (4). Put back four screws and starwashers (5).
4. Unscrew two nuts (6) and take off oil inlet tube (7).
5. Take out two screws and lockwashers (8) and take off oil inlet tube adapter (9). Take off and throw away adapter gasket (10) .

GO TO FRAME 4

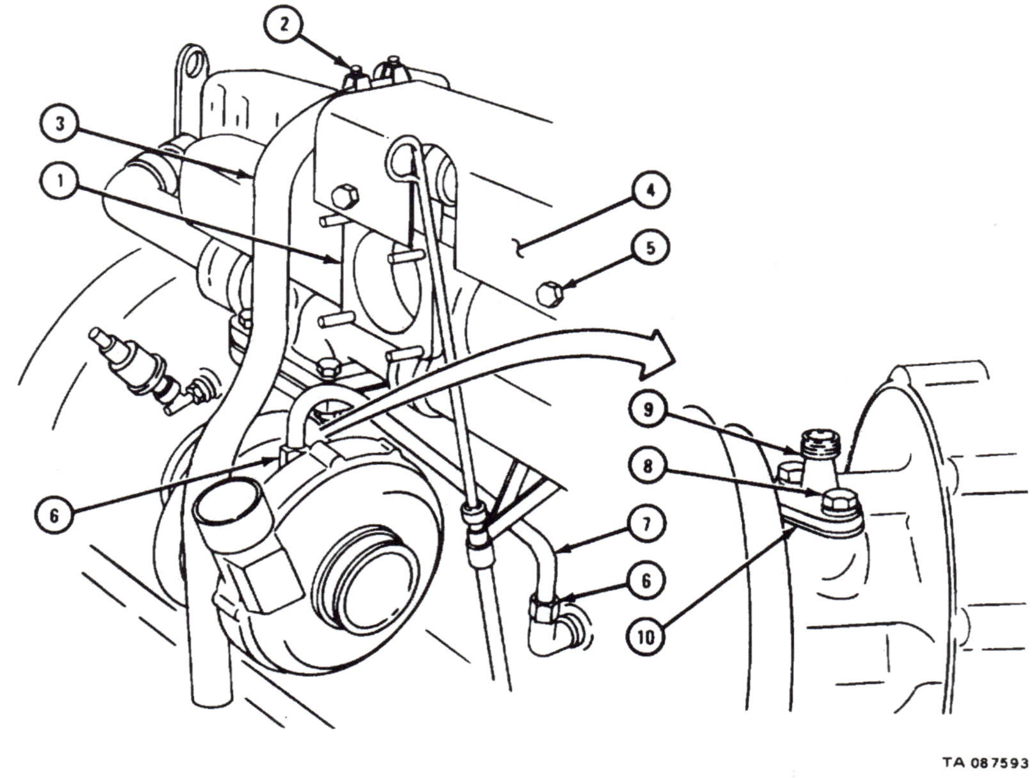

FRAME 4

Soldier A 1. Hold up turbocharger (1) so it will not fall when soldier B takes out nuts (2) and screws (3).

Soldier B 2. Take out four nuts and locknuts (2) and two screws (3).

Soldier A 3. Take out turbocharger (1). Take off and throw away turbocharger gasket (4).

4. Plug or tape all openings in turbocharger (1).

GO TO FRAME 5

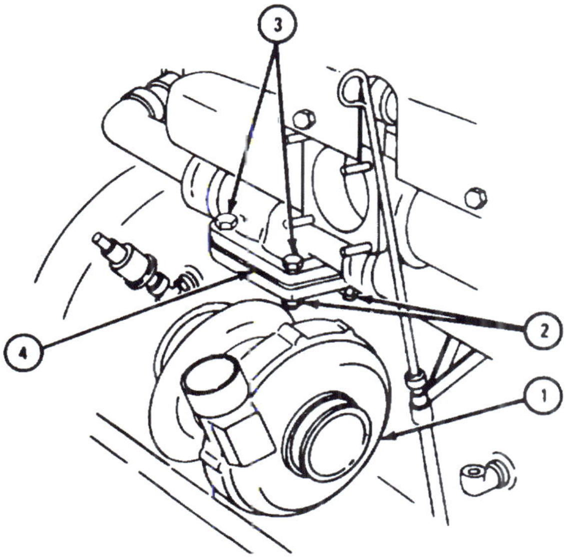

TA 087594

FRAME 5

1. Unscrew and takeout sparkplug (1). Take off and throw away spark plug gasket (2) .

CAUTION

Do not use open end wrench to take out fuel inlet adapter (3) or fuel return adapter (4). Adapter s may be bent. Use only tubing wrench or box wrench.

2. Unscrew and take out fuel inlet adapter (3) and fuel return adapter (4).

3. Loosen locknut (5). Unscrew and take out nozzle and check valve assembly (6).

END OF TASK

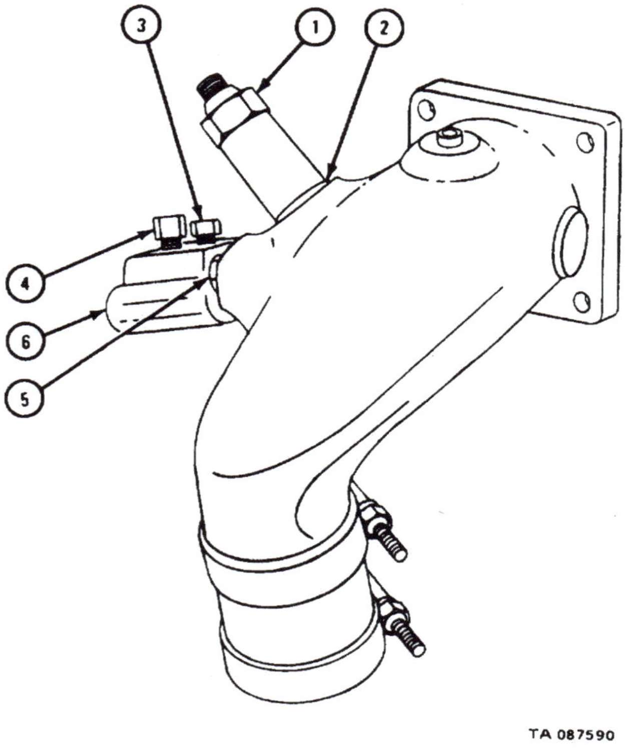

b. Engine LDS-465-1 .

FRAME 1

1. Loosen hose clamp (1).
2. Take out screw (2), washer (3), lockwasher (4), and nut (5) .
3. Take out crankcase breather tube (6) .
4. Bend back locktabs and take out six screws (7) and three lockplates (8).
5. Take out three screws and lockwashers (9) .
6. Take off exhaust elbow with bracket and inner gasket (10). Throw away gasket.

GO TO FRAME 2

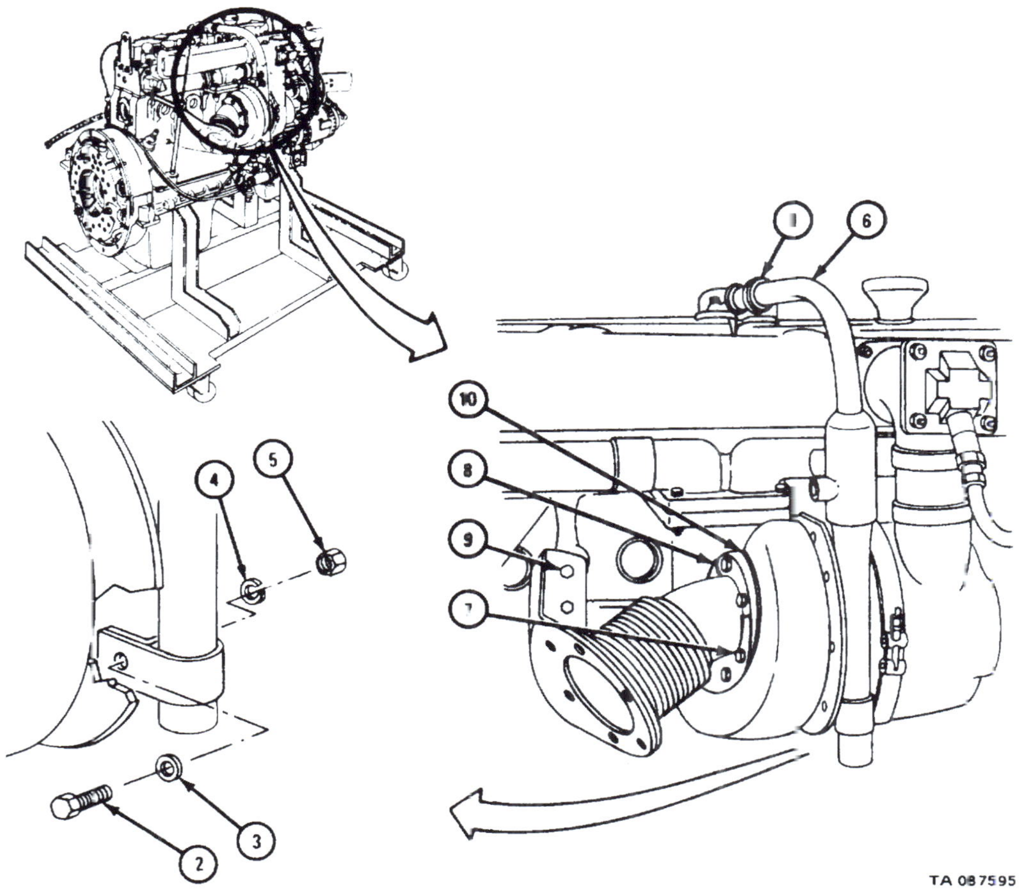

FRAME 2

1. Loosen lower hose clamp (1).

2. Loosen two hose clamps (2) and slide oil drain hose (3) all the way up on oil drain tube (4).

3. Take out two screws and lockwashers (5).

Soldier A 4. Hold up turbocharger (6) so it does not fall when locknuts (7) and screws (8) are taken off.

Soldier B 5. Take off four locknuts (7) and two screws (8).

Soldier A 6. Take out turbocharger (6).

7. Take off and throw away two gaskets (9).

NOTE

Hose (10) should stay on engine.

8. Tighten hose clamp (1).

GO TO FRAME 3

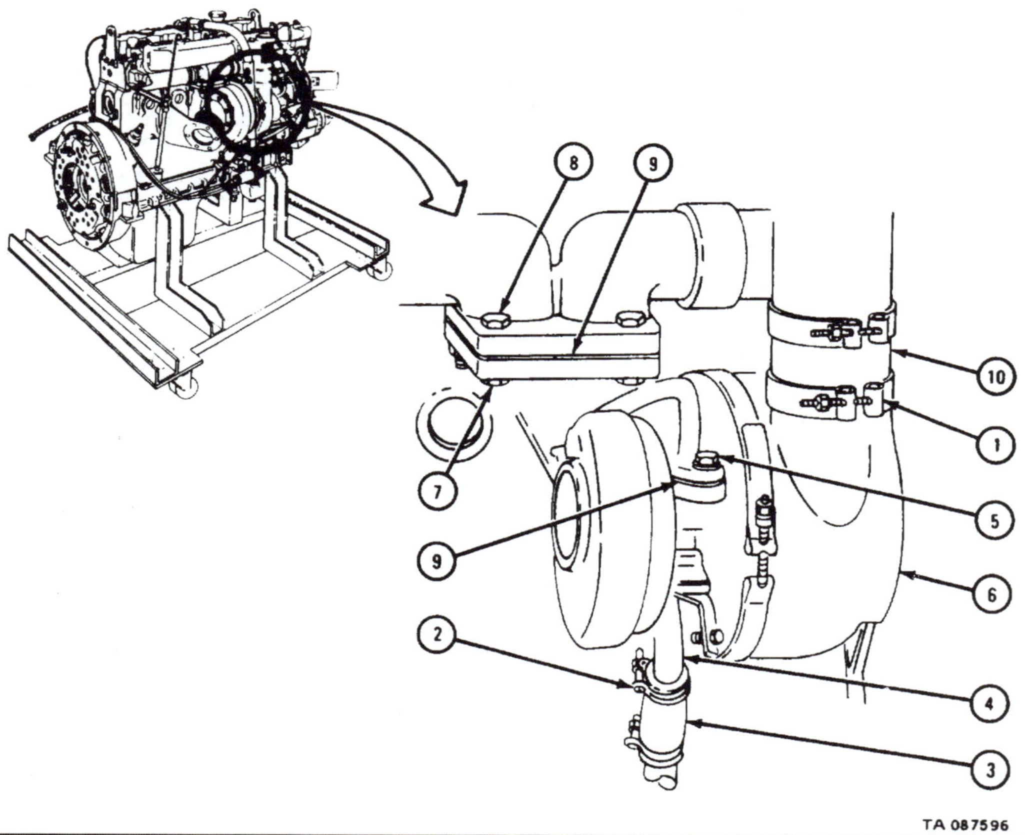

FRAME 3

1. Unscrew nut (1) and take off oil inlet tube adapter (2).
2. Unscrew nut (3) and take out turbocharger oil inlet tube (4).
3. Unscrew and take out oil drain hose adapter (5).

GO TO FRAME 4

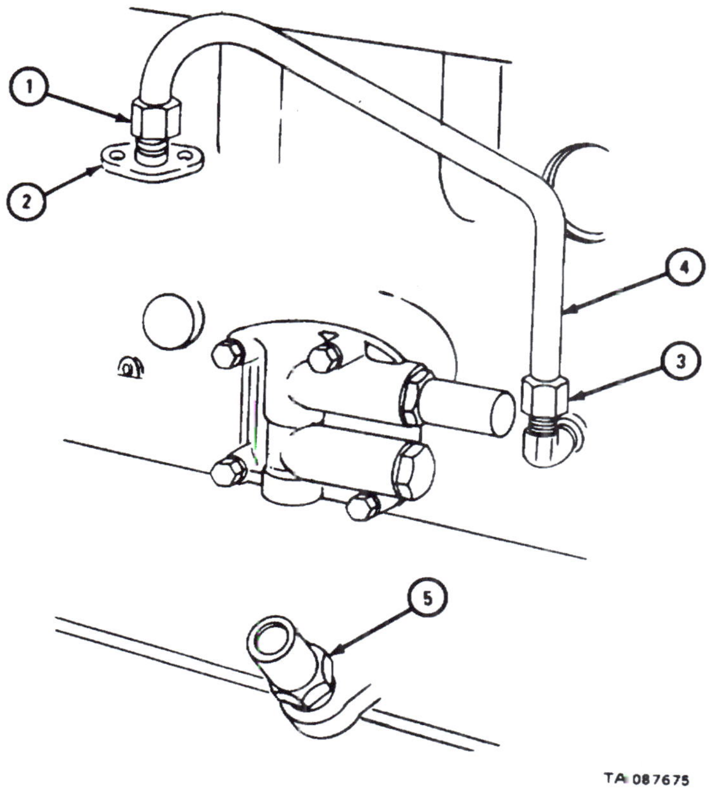

FRAME 4

1. Take off and tag hose with clamps (1).
2. Take out two screws and lockwashers (2) and take off oil drain tube (3).
3. Take off gasket (4) and throw it away.

CAUTION

Plug or tape all air and oil openings on turbocharger (5) to keep out dirt. Dirt can damage parts inside turbocharger.

END OF TASK

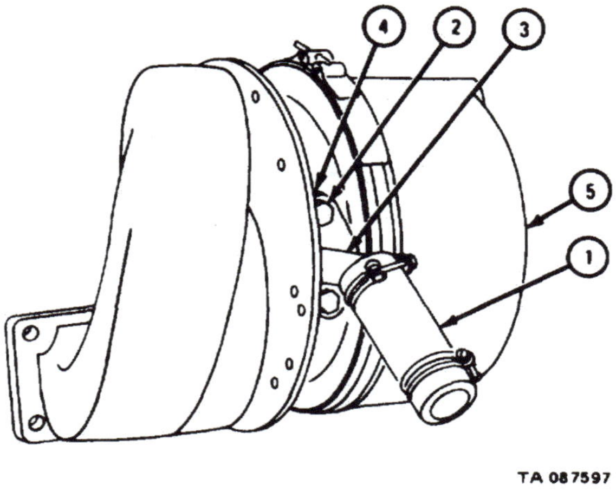

c. Engines LDS-465-1A and LDS-465-2.

NOTE

If working on engine LDS-465-2, go to frame 2. Turbocharger may already have been taken off engine LDS-465-1A when engine was taken out of truck .

FRAME 1

1. Loosen hose clamp (1).
2. Take off nut (2), lockwasher (3), flat washer (4), and screw (5) .
3. Take out breather tube (6) with clamp (7).

GO TO FRAME 2

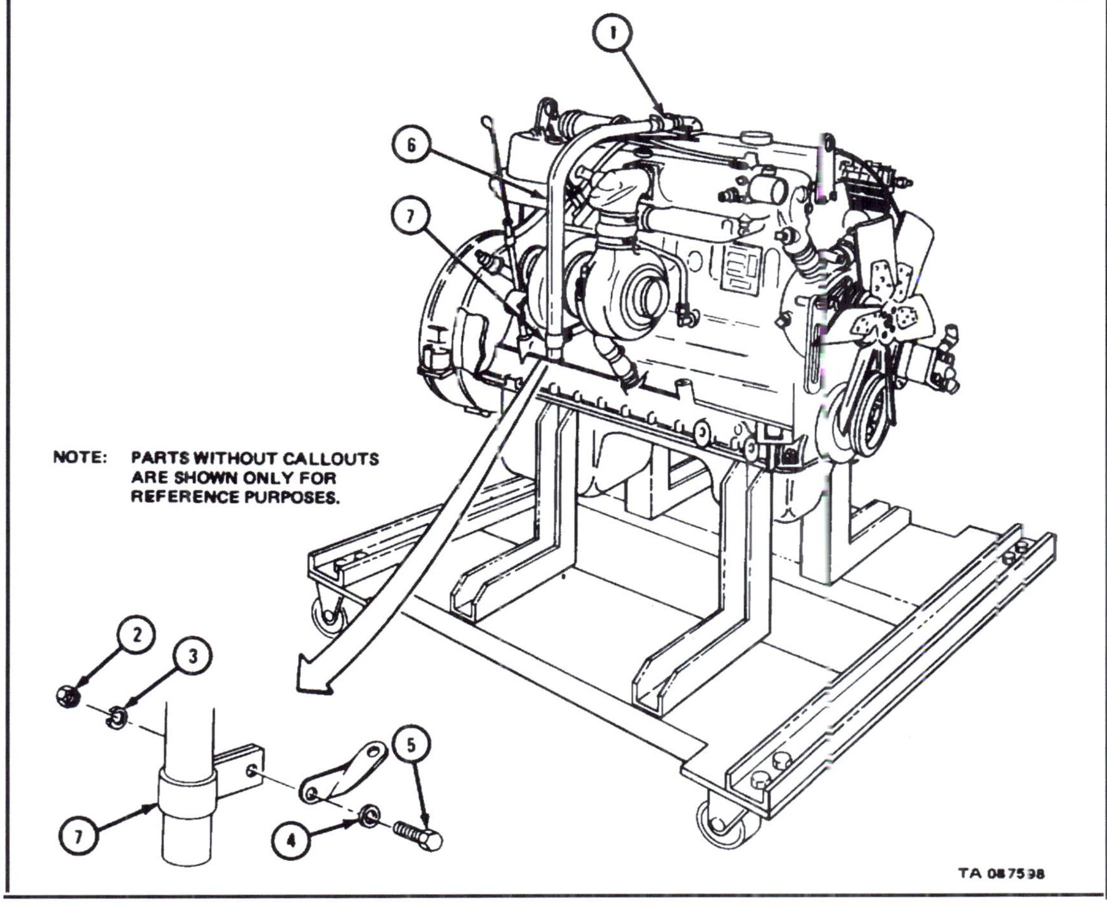

FRAME 2

1. Loosen two hose clamps (1). Slide clamps to center of oil drain hose (2).
2. Take out two screws and lockwashers (3).
3. If working on engine LDS-465-1A, take out bracket (4) .
4. Pull out oil drain tube (5). Take off and throw away oil drain tube gasket (6). Take off and tag oil drain hose (2).
5. Unscrew and take out oil drain hose adapter (7).
6. Loosen but do not take off hose clamp (8). Do not loosen hose clamp (9).

GO TO FRAME 3

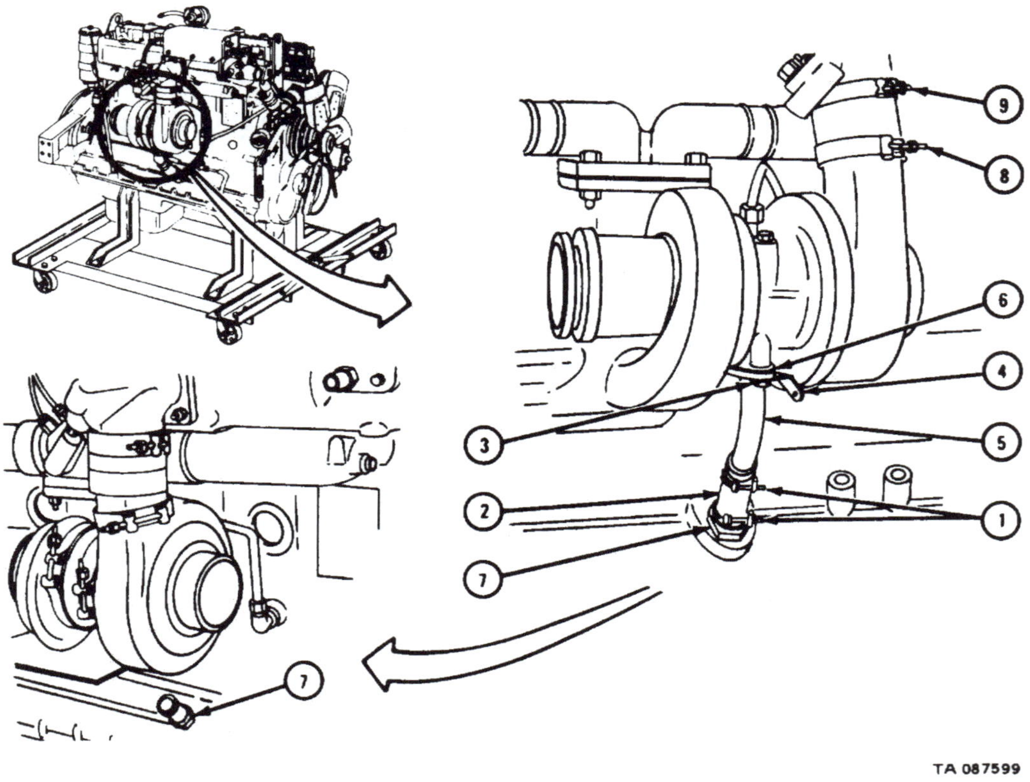

FRAME 3

1. Unscrew nut (1).
2. Unscrew nut (2) and take out oil inlet tube (3).

CAUTION

Plug or tape over oil inlet tube opening on engine to keep out dirt. Dirt can damage parts inside engine.

3. Take out two screws and lockwashers (4) and take off oil inlet adapter (5). Take off oil inlet adapter gasket (6) and throw it away.

Soldier A 4. Hold up turbocharger (7) so it does not fall when locknuts (8) and screws (9) are taken off.

Soldier B 5. Take out four locknuts (8) and two screws (9).

Soldier A 6. Take out turbocharger (7). Take off and throw away turbocharger gasket (10).

CAUTION

Plug or tape over all air and oil openings on turbocharger (7) to keep out dirt. Dirt can damage parts inside turbocharger.

GO TO FRAME 4

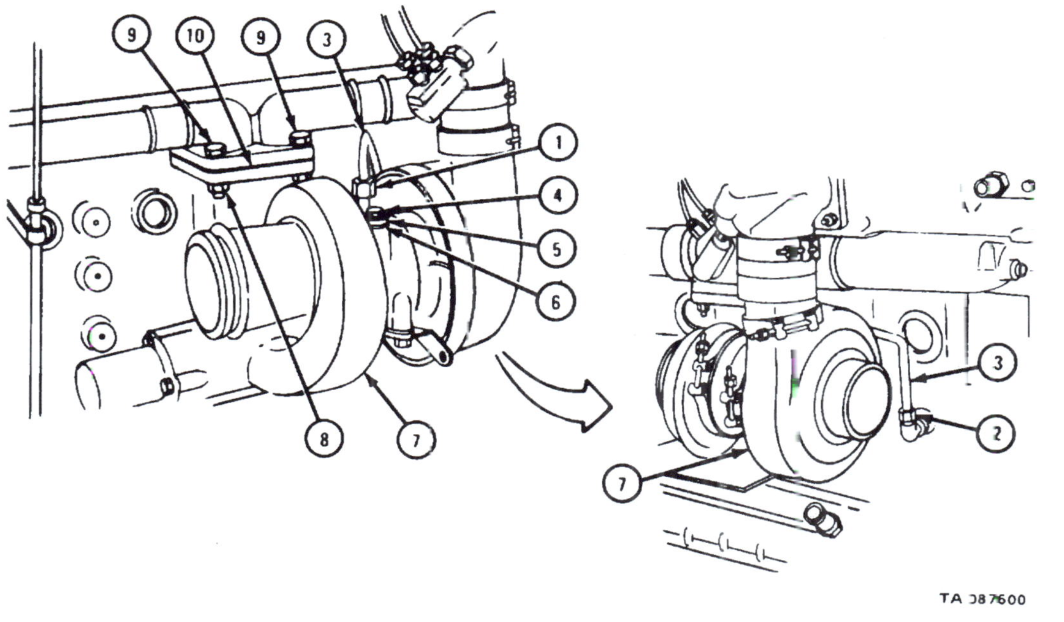

FRAME 4

NOTE

Not all engines have baffle plate (3) .

1. Take out screws and lockwashers (1 and 2). Take off baffle plate (3).

2. Put back screws and lockwashers (2).

END OF TASK

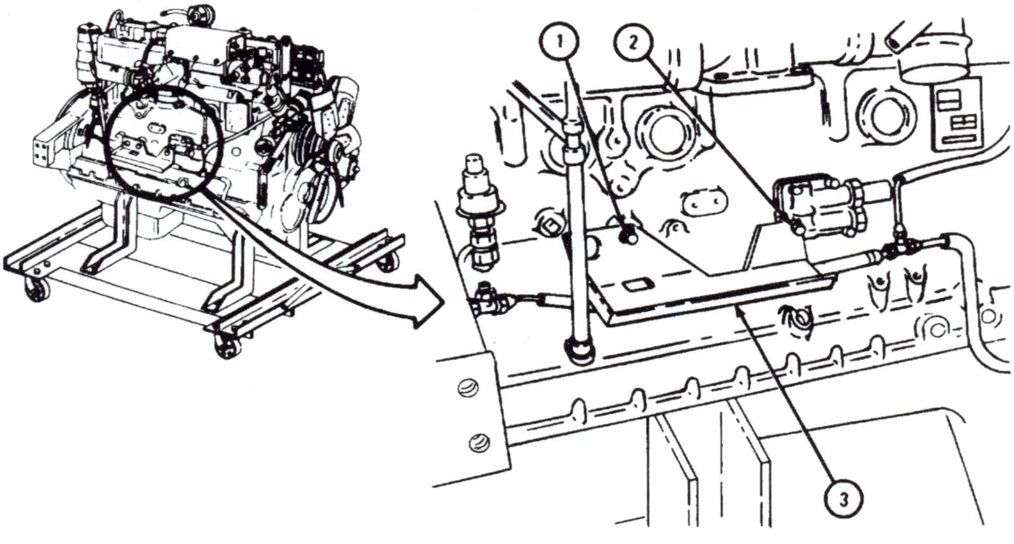

3-6. CRANKCASE BREATHER TUBE (ENGINES LD-465-1 AND LD-465-1C).

FRAME 1

NOTE

Engine shown is LD-465-1C. This task is the same for engine LD-465-1.

1. Loosen hose clamp (1).
2. Take out screw (2), washer (3), lockwasher (4), and nut (5) and take out crankcase breather tube (6) with clamp (7).

END OF TASK

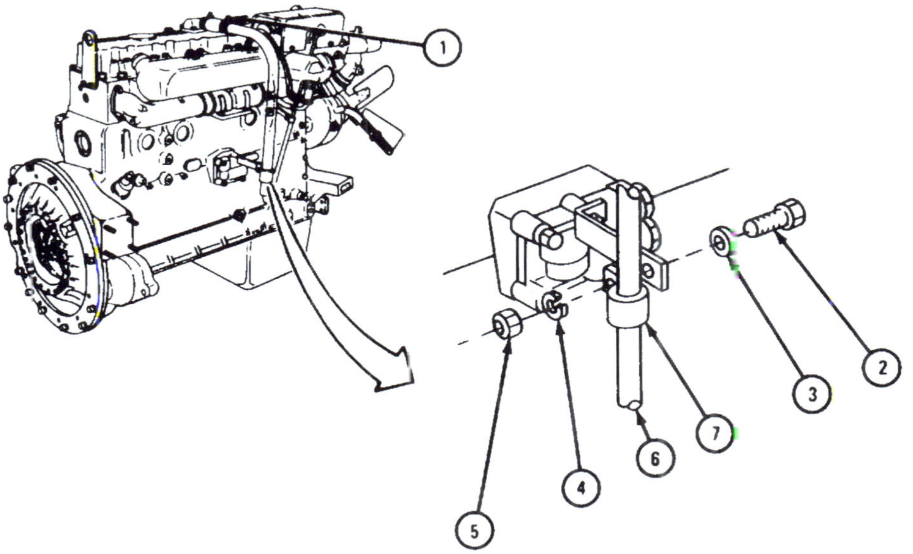

TA 086446

3-7. DISCONNECTION OF FUEL INJECTION PUMP FUEL LINES AND OIL HOSES.

a. Engines LD-465-1, LD-465-1C, and LDT-465-1C .

FRAME 1

1. Unscrew nut (1) . Takeoff and ta g pressure oil hose (2) .
2. Unscrew nut (3) . Takeoff and ta g fuel injector nozzles-to-fuel pump return tube (4) .
3. Unscrew nut (5) . Takeoff and ta g fuel pump-to-fuel filter return tube (6) .
4. Unscrew three nuts (7). Take off and tag fuel pump-to-fuel filter tube (8), fuel pump-to-flame heater tube (9), and fuel filter-to-fuel density compensator tube (10) .
5. Plug or cap all openings in fuel injection pump assembly.

END OF TASK

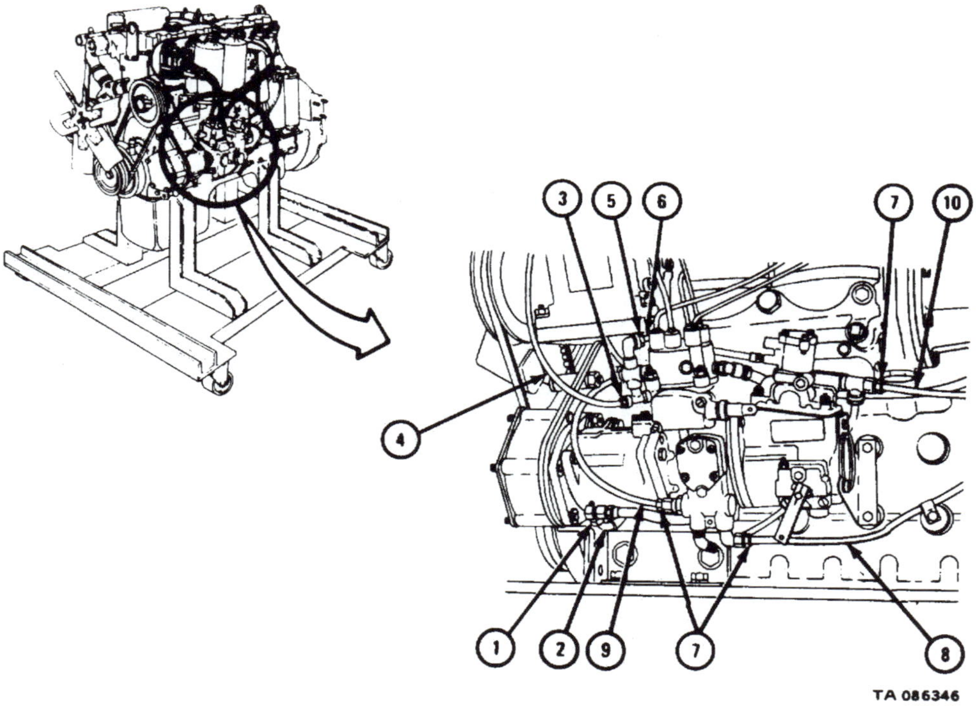

b. Engine LDS-465-1A.

FRAME 1

1. Unscrew nut (1). Takeoff and tag pressure oil hose (2)
2. Unscrew nut (3). Take off and tag fuel injector nozzle-to- fuel pump return tube (4).
3. Unscrew nut (5). Takeoff and tag fuel pump-to-fuel filter return tube (6).
4. Unscrew three nuts (7). Take off and tag fuel pump-to-fuel filter tube (8), fuel pump-to-flame heater tube (9), and fuel filter-to-fuel density compensator tube (10).
5. Plug or cap all openings in fuel injection pump assembly.

END OF TASK

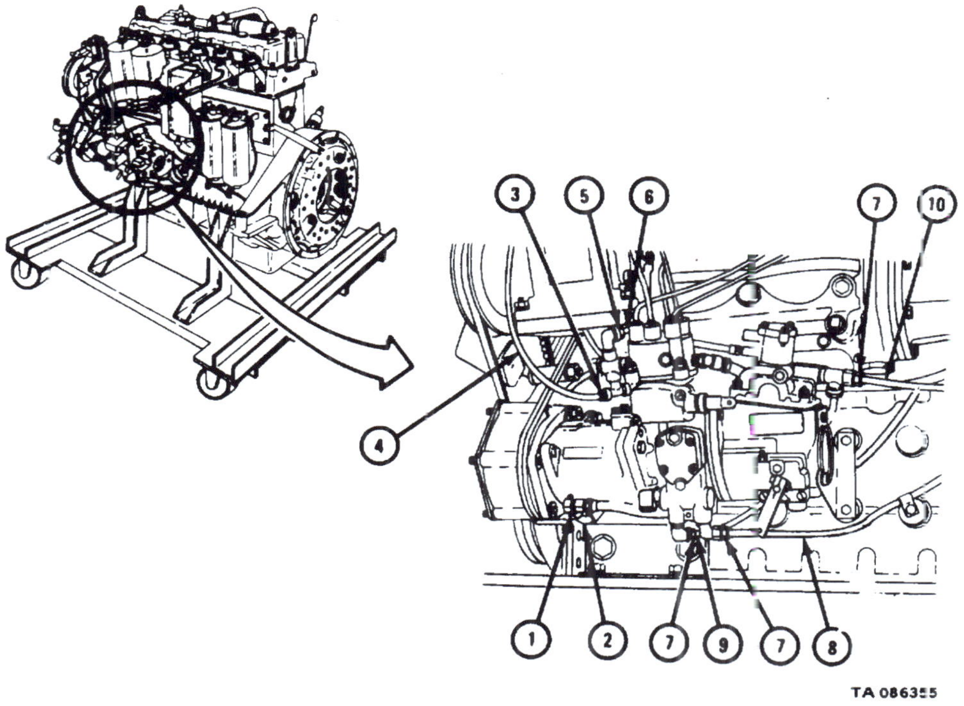

c. Engine LDS-465-1 .

FRAME 1

1. Unscrew nut (1). Take off and tag pressure oil hose (2) .
2. Unscrew four nuts (3). Take off and tag fuel injector nozzle-to-fuel pump return tube (4), fuel pump-to-fuel filter return tube (5), fuel pump-to-flame heater tube (6), and fuel pump-to-fuel filter tube (7).
3. Unscrew nut (8). Take off and tag fuel filter-to-fuel density compensator tube (9) .
4. Plug or cap all openings in fuel injection pump assembly.

IF WORKING ON ENGINE WITH OIL DRAIN HOSE (10), GO TO FRAME 2.
IF WORKING ON ENGINE WITHOUT OIL DRAIN HOSE (10), END OF TASK

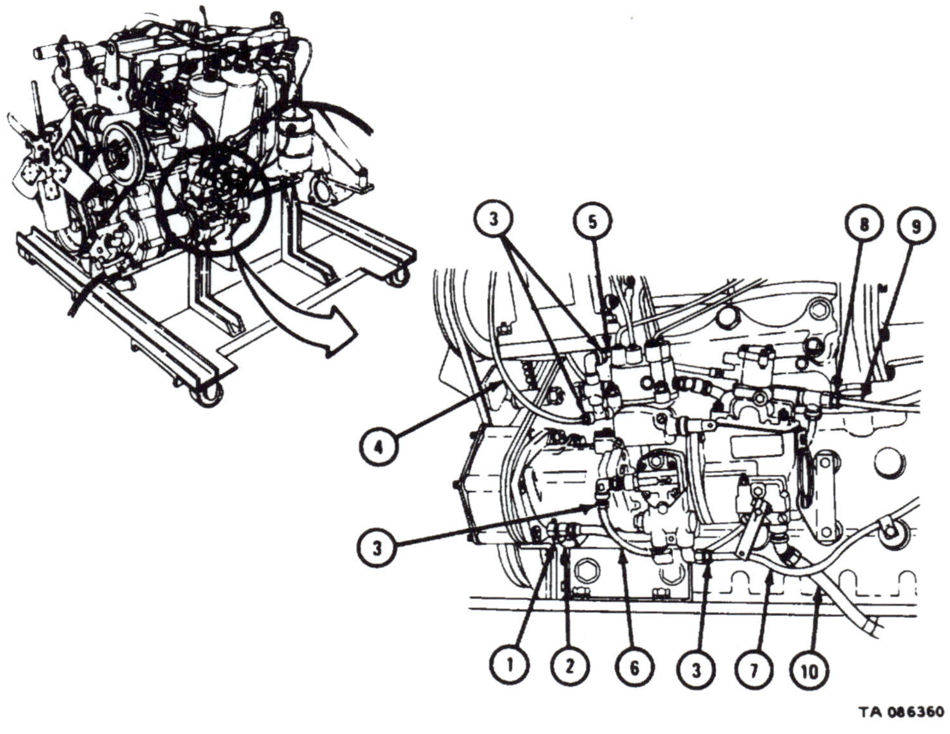

FRAME 2

1. Unscrew nut (1) and takeoff oil drain hose (2).
2. Takeout screw and lockwasher (3) and takeoff hose clamp (4). Put back screw and lockwasher (3).
3. Unscrew nut (5). Take out and tag oil drain hose (2).
4. Unscrew and take out pipe elbow (6).
5. Plug or cap drain hose in fuel injection pump assembly (7).

END OF TASK

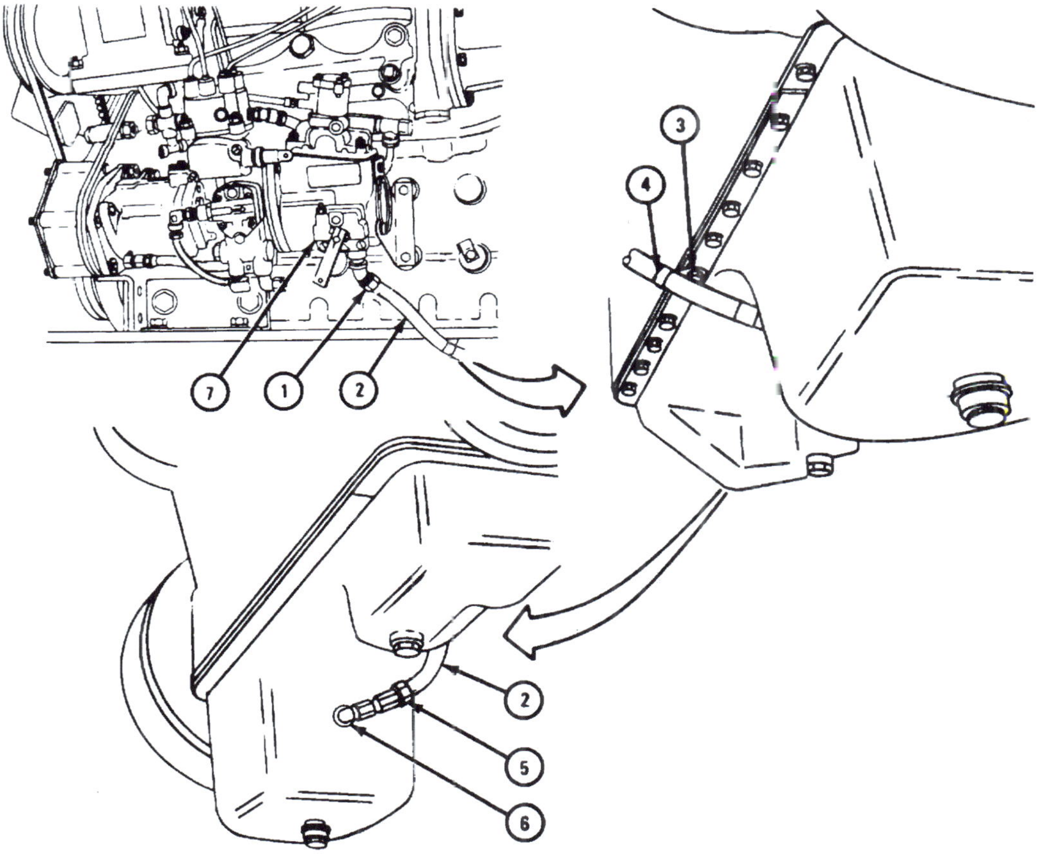

d. Engine LDS-465-2 .

FRAME 1

NOTE

Tag all tubes and hoses so they are put back in the same place.

1. Unscrew nut (1) and take off pressure oil hose (2).
2. Unscrew nut (3) and take off fuel injector nozzle-to-fuel pump return tube (4).
3. Unscrew nut (5) and take off fuel pump-to-flame heater tube (6).
4. Unscrew nut (7) and take off fuel pump-to-fuel filter tube (8).
5. Unscrew nut (9) and take off fuel filter-to-fuel density compensator tube (10).
6. Plug or cap all openings in fuel injection pump assembly.

END OF TASK

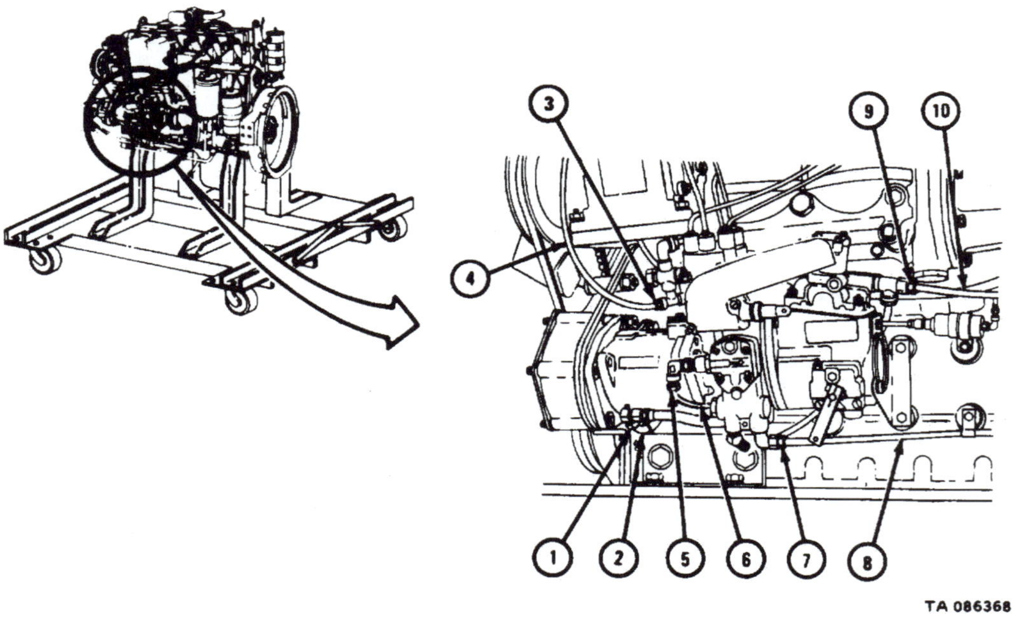

3-8. FUEL INJECTOR TUBES.

a. Tube Clamps.

(1) All engines except LDS-465-2.

FRAME 1

NOTE

Some engines have six tube clamps (2) and some have two tube clamps. Procedure shown is for engines with six clamps.

1. Take out four screws and locknuts (1) and take off four tube clamps (2).
2. Take out two screws, four flat washers, and two locknuts (3) and take off outer half of tube clamp (4).
3. Take off nut, lockwasher, and flat washer (5) and take off inner half of tube clamp (4).
4. Take out two screws, flat washers, and locknuts (6) and take off tube clamp (7).

NOTE

Some engines do not have bracket (9).

5. Take out screw (8) and take off bracket (9).

NOTE

Some engines do not have lockbolts (10).

6. Take out two lockbolts (10).

END OF TASK

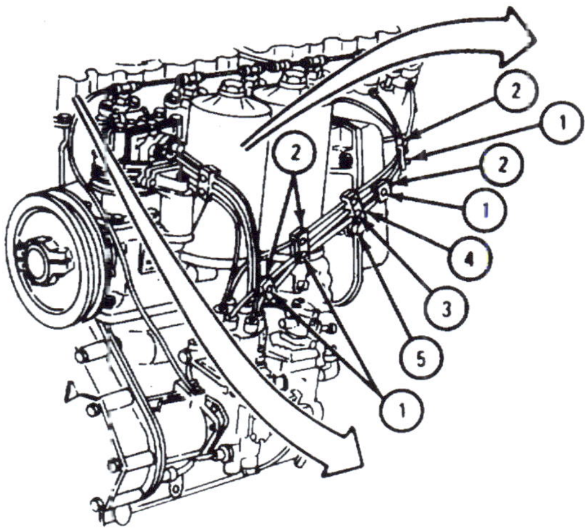

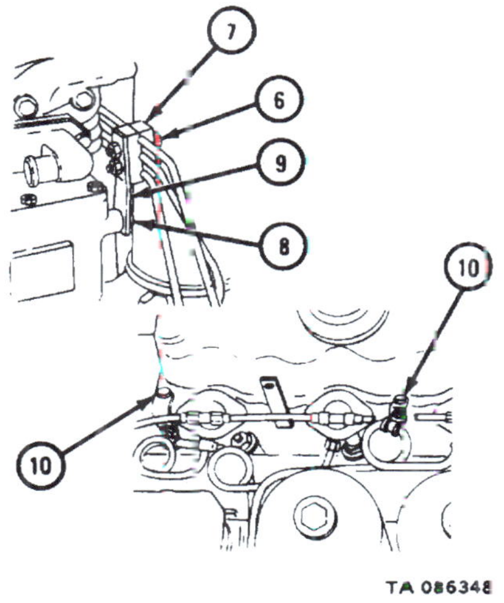

(2) Engine LDS-465-2 .

FRAME 1

1. Take off six locknuts, flat washers, and screws (1) and take off three tube clamps (2) .
2. Take off two locknuts and screws (3) and take off two tube clamps (4).
3. Take off two screws (5), and nuts and flat washers (6), and take off clamp (7).
4. Take out screw (8) and take off bracket (9).

END OF TASK

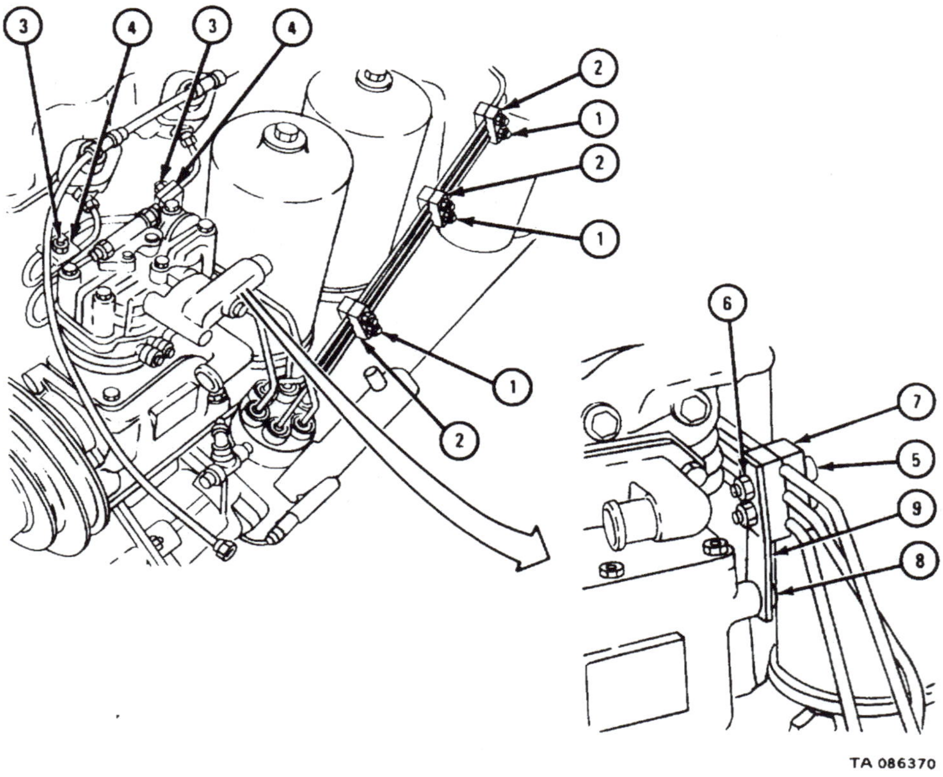

b. Fuel Injector Tubes.

FRAME 1

NOTE

Tag six fuel injector tubes with numbered tool tags so that they are put back in the same place.

1. Slide six dust caps (1) up fuel injector tubes (2).
2. Unscrew six fuel injector tube nuts (3) and six fuel injector tube nuts (4). Take out six fuel injector tubes (2).
3. Plug six holes in fuel injection pump head (5) and plug hole in each of six fuel injector nozzles (6).

END OF TASK

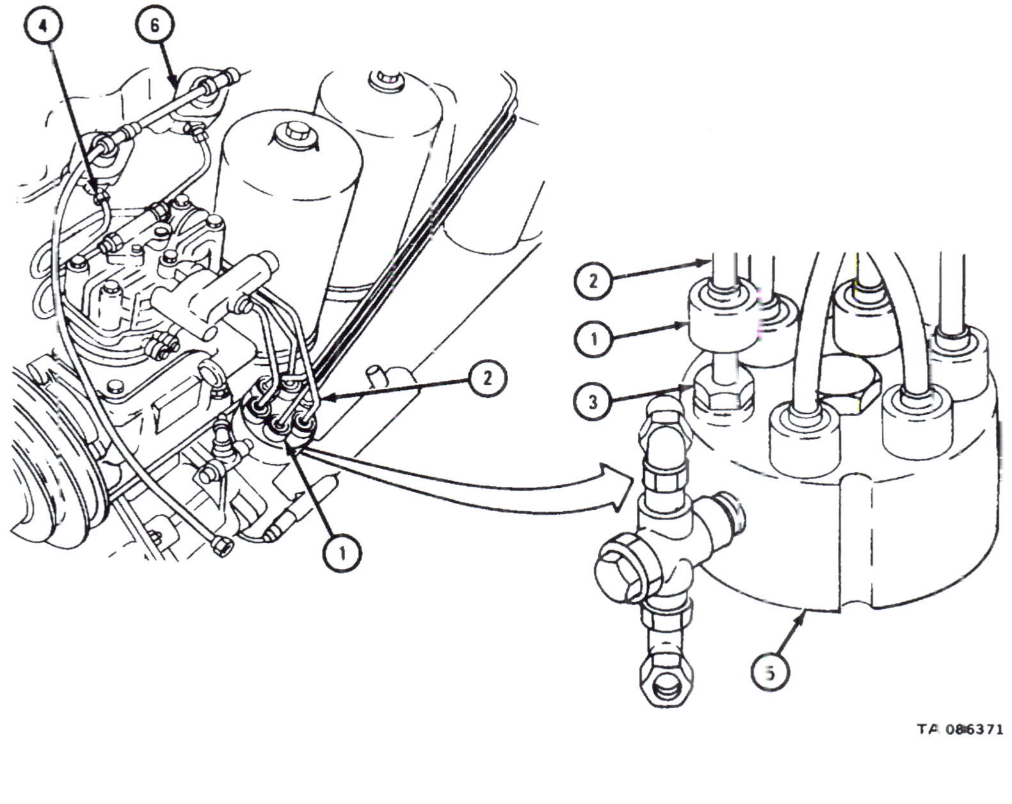

3-9. FUEL INJECTION PUMP DRIVE GEAR ACCESS COVER OR HYDRAULIC PUMP ASSEMBLY.

a. Engines LD-465-1, LD-465-1C, and LDT-465-1C .

FRAME 1

1. Take off three nuts and lockwashers (1) .
2. Take off nut and lockwasher (2) and screw (3) .
3. Take off nut and lockwasher (4) and screw (5).
4. Take off access cover (6) . Take off and throw away access cover gasket (7).

END OF TASK

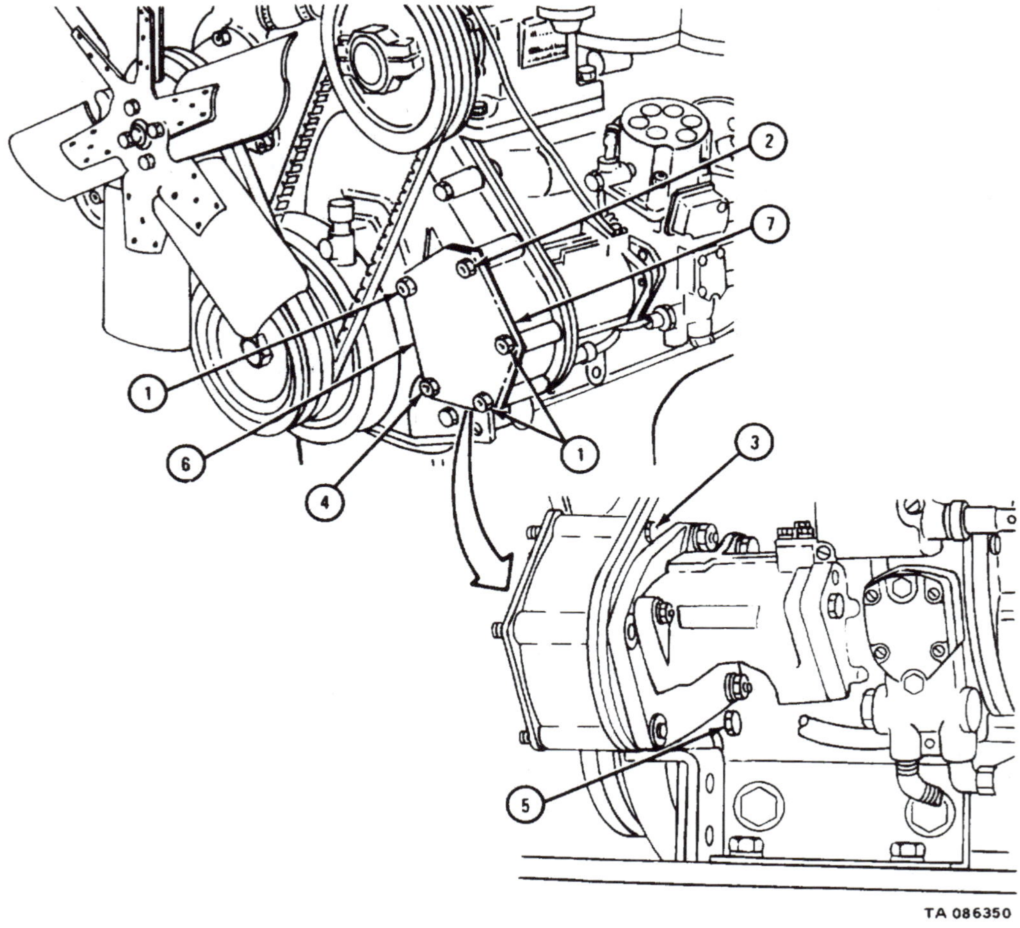

b. <u>Engines LDS-465-1, LDS-465-1A, and LDS-465-2</u>.

NOTE

If working on engine LDS-465-1 or LDS-465-1A, go to frame 2.

FRAME 1

NOTE

Tag hydraulic pump return hose (1) and hydraulic pump bypass hose (2) so that they are put back in the same place.

1. Loosen two hose clamps (3).
2. Place container under hydraulic pump return hose (1). Take off hydraulic pump return hose and drain hydraulic oil into container.
3. Place container under hydraulic pump bypass hose (2). Take off hydraulic pump bypass hose and drain hydraulic oil into container.
4. Empty container in approved disposal area.

GO TO FRAME 2

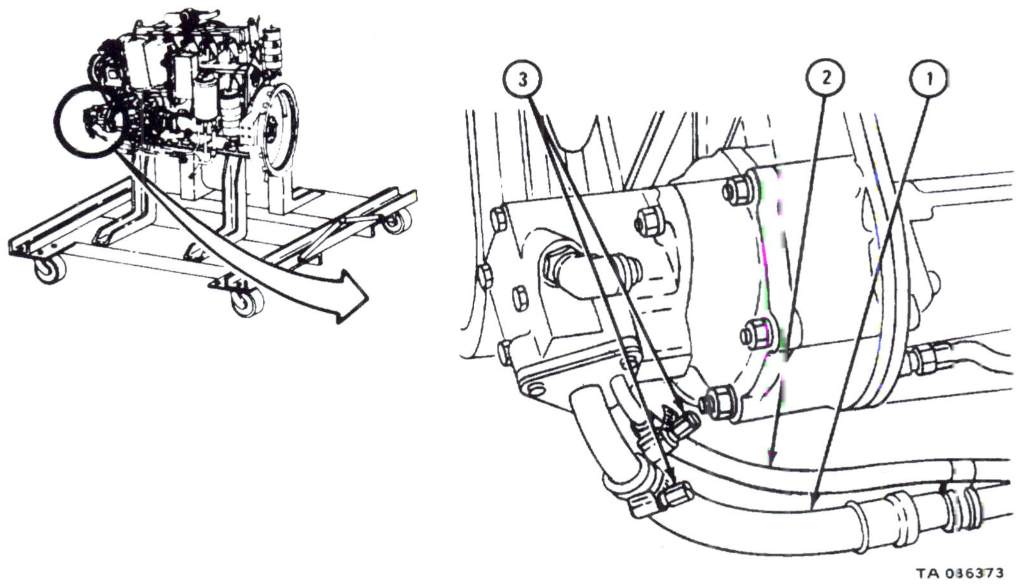

FRAME 2

NOTE

Hydraulic pump shown is for engines LDS-465-1 and LDS-465-1A. This task is the same for engine LDS-465-2.

1. Take off two nuts and lockwashers (1).
2. Take off nut and lockwasher (2) and screw (3) .

GO TO FRAME 3

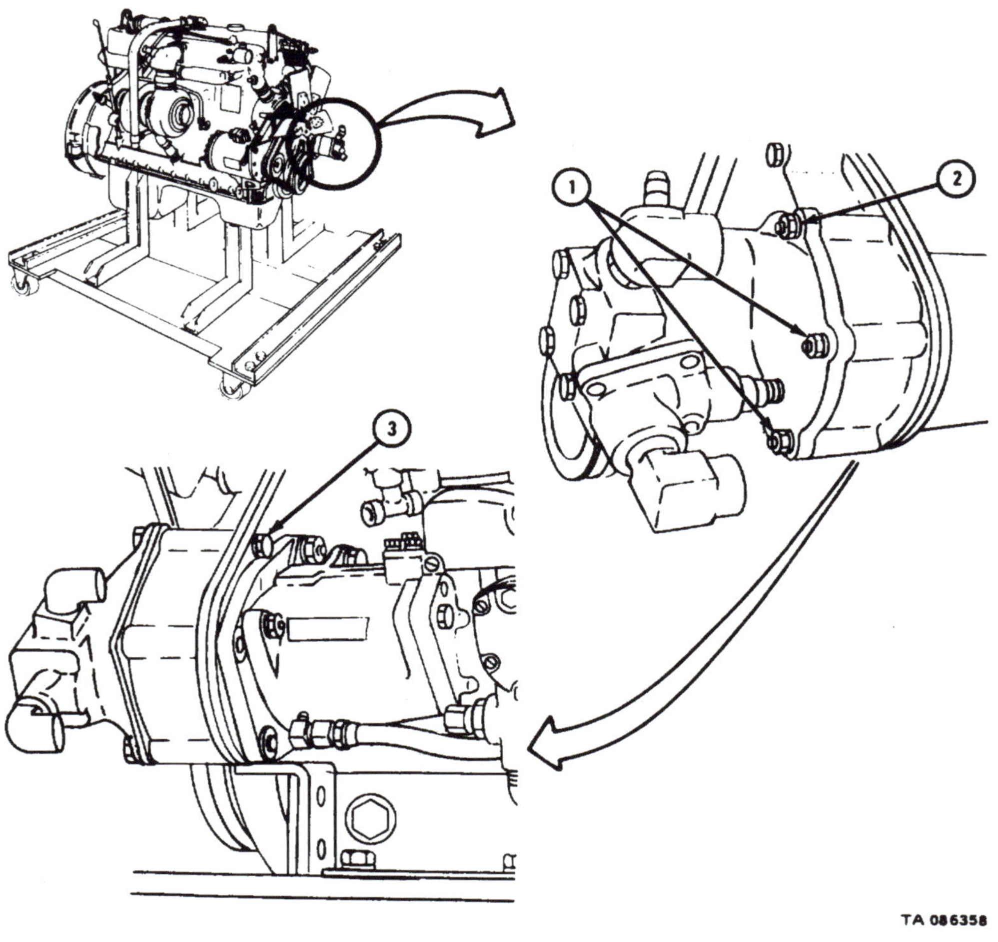

FRAME 3

1. Takeoff nut and lockwasher (1) .

2. Take off nut and lockwasher (2) and screw (3), and take off hydraulic pump (4) with mounting adapter (5) .

3. Take off hydraulic pump mounting adapter gasket (6) and throw it away.

END OF TASK

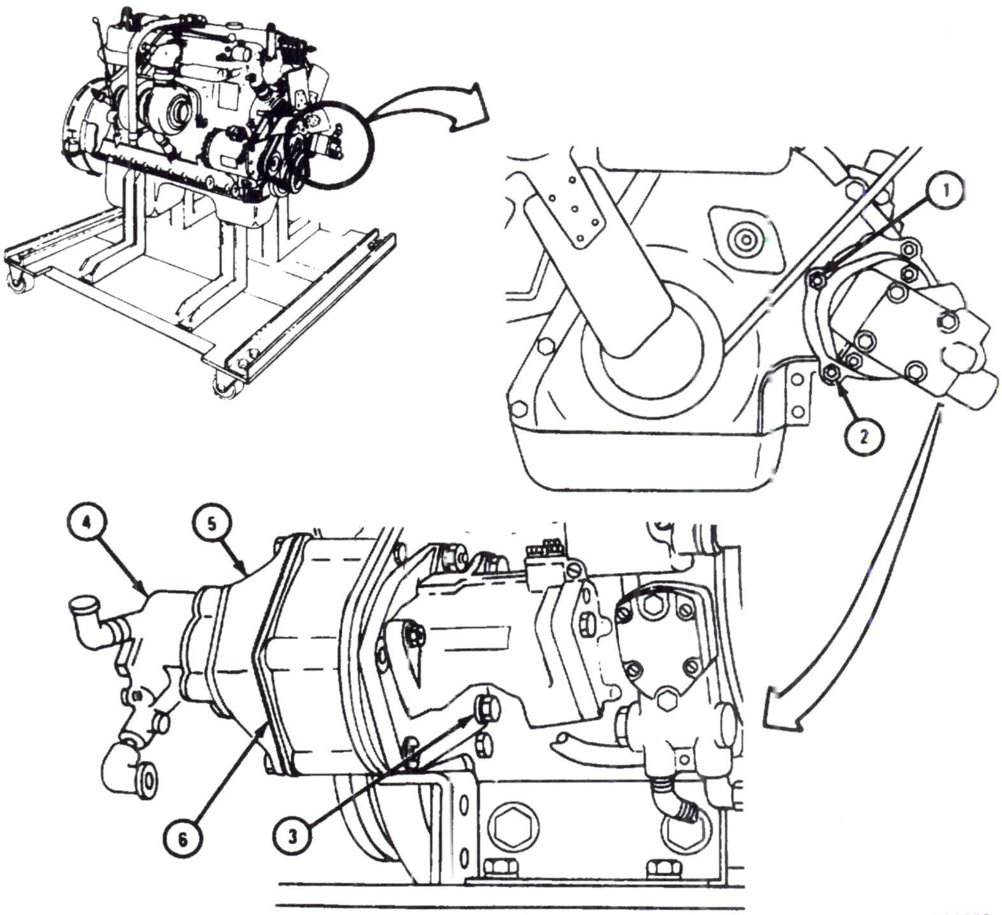

TA 086359

3-10. FUEL INJECTION PUMP ASSEMBLY.

FRAME 1

1. Takeout three screws and lockwashers (1) .
2. Take off retaining plate (2).
3. Takeout fuel pump drive gear (3) .

GO TO FRAME 2

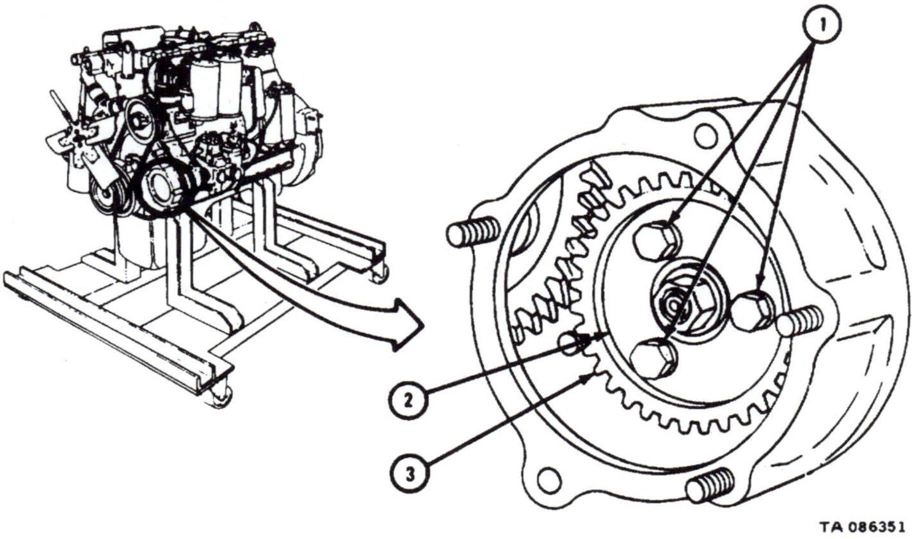

FRAME 2

NOTE

Fuel injection pump shown is for engines LD-465-1, LD-465-1C, and LDT-465-1C. This task is the same for other engines . Some engines do not have lockplate (1) .

1. Bend back locktabs on lockplate (1). Take out two screws (2) and lockplate (1) .
2. Take out two screws and lockwashers (3) and take off bracket (4).

GO TO FRAME 3

FRAME 3

Soldier A 1. Hold up fuel pump assembly (1) so it does not fall when mounting screws are taken out.

Soldier B 2. If fuel pump (1) is held with screw (2) as in view A, unscrew but do not take out screw.

3. If fuel pump is held with nut and lockwasher (3) on stud (4) as in view B, take off nut and lockwasher (3).

4. Take out two screws and lockwashers (5) .

Soldier A 5. Pull fuel pump assembly (1) toward back of engine and take it out. If engine has mounting stud (4), make sure that pump assembly clears it.

6. Put container under front end of fuel pump assembly (1). Tilt fuel pump forward and drain oil inside fuel pump into can.

NOTE

Preformed packing (6) may stick to fuel injection pump assembly (1).

7. Take off and throw away preformed packing (6).

GO TO FRAME 4

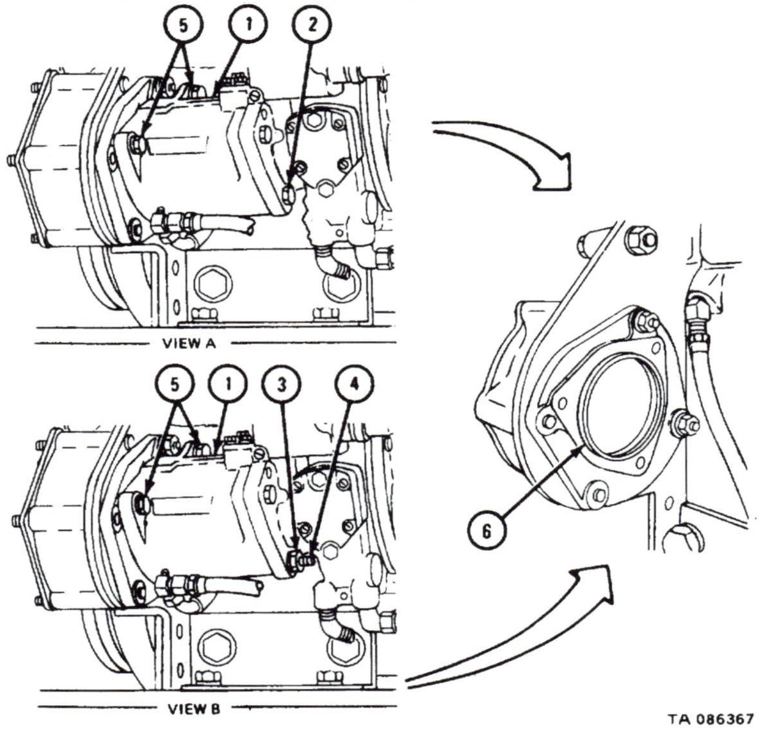

FRAME 4

NOTE

Fuel injection pump (1) shown is for engines LD-465-, LD-465-1C, and LDT-465-1C. This task is the same for other engines.

Have machine shop make holding wrench (2). See figure 3-1 .

1. Join holding wrench (2) to fuel injection pump drive gear hub (3) using two screws (4) as shown .
2. Take off nut (5) .
3. Take out two screws (4) and take off holding wrench (2).
4. Pull off drive gear hub (3) .
5. Put nut (5) back on fuel pump driveshaft (6).

END OF TASK

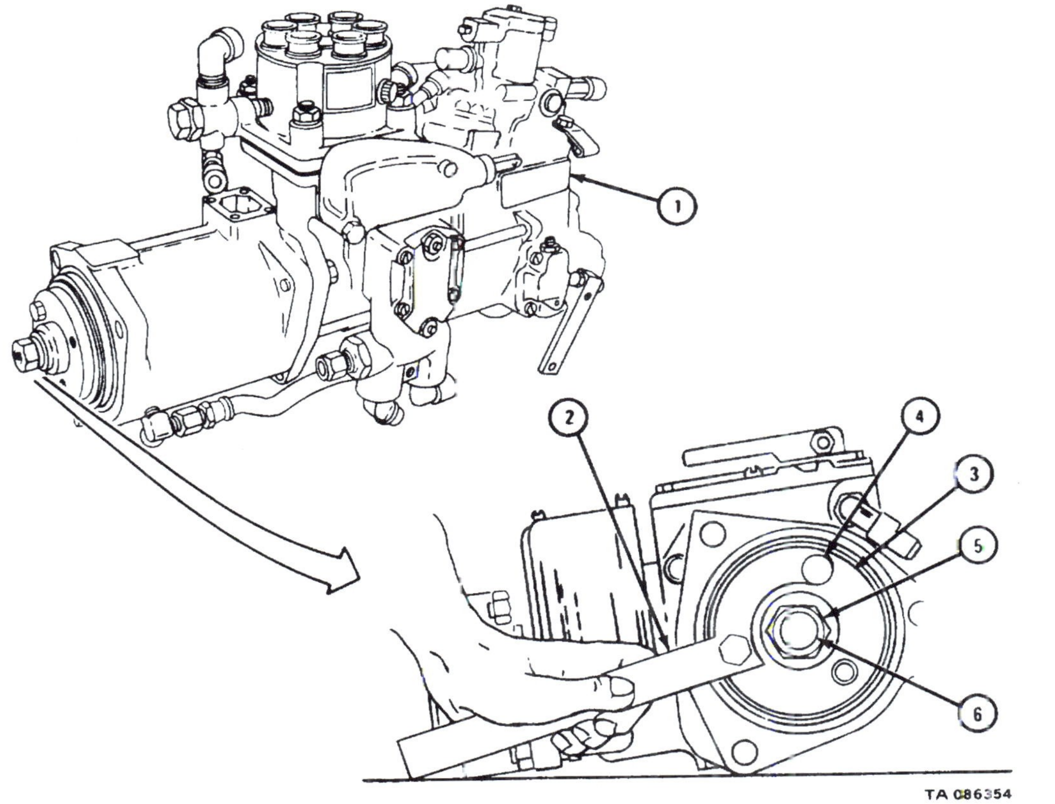

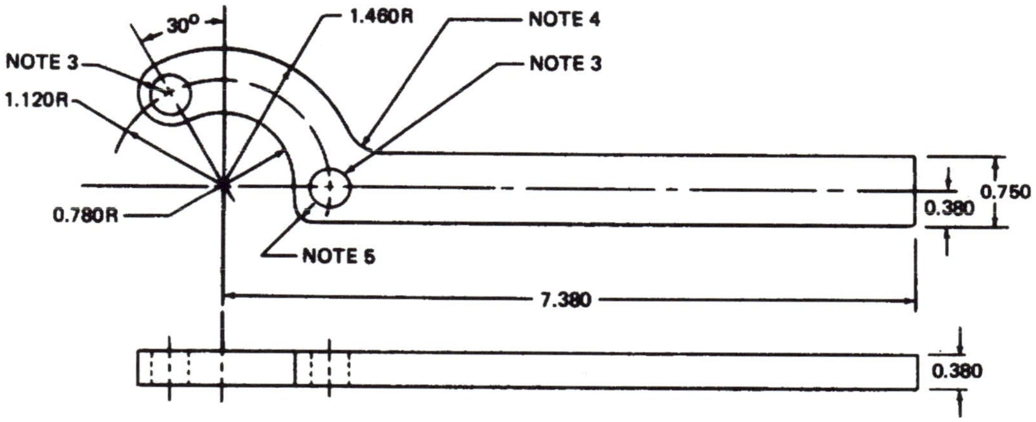

NOTES:
1. ALL DIMENSIONS GIVEN ARE IN INCHES
2. MATERIAL: COLD ROLL STEEL
3. DRILL TWO HOLES 0.406 IN.
4. RADIUS 0.750 IN.
5. RADIUS 0.500 IN.

TA 121159

Figure 3-1. Fuel Injection Pump Gear Hub Holding Wrench Fabrication Instructions

3-11. OIL FILTER ELEMENTS.

FRAME 1

1. Hold container under oil filter drain plug (1). Takeout drain plug, and drain oil out of oil filter housing (2). Put back drain plug (1).

NOTE

There is a spring and spring retainer cup inside oil filter bodies (3 and 4). Be careful they are not lost when filter bodies are taken off.

2. Unscrew center post (5) . Lift out filter body (3) with center post.
3. Take out and throw away filter element (6).
4. Take out and throw away filter body gasket (7).
5. Do steps 2 through 4 again for front oil filter body (4).

END OF TASK

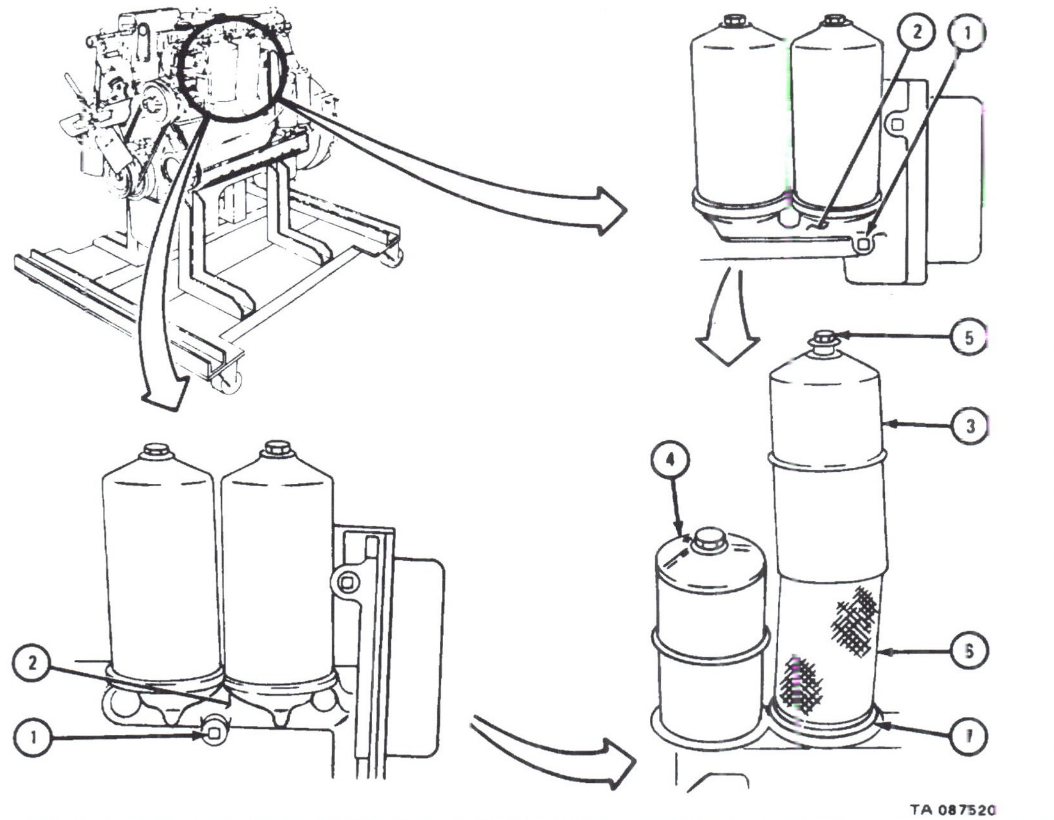

3-12. AIR COMPRESSOR ASSEMBLY,

a. All Engines Except LDS-465-2.

FRAME 1

1. Takeout two screws and lockwashers (1) .
2. Place air compressor pulley adjusting wrench (2) over air compressor pulley flange (3) as shown .
3. Turn pulley adjusting wrench (2) to left . Take off and tag air compressor drive belt (4) .
4. Turn pulley flange (3) to the right and put in two screws and lockwashers (1) so they are not lost.
5. Take out screw (5) and take off clip (6) with fuel return tube (7). Put back screw (5) .
6. Take off four nuts (8) and lockwashers (9) and lift off air compressor (10). Take off and throw away air compressor gasket (11).

END OF TASK

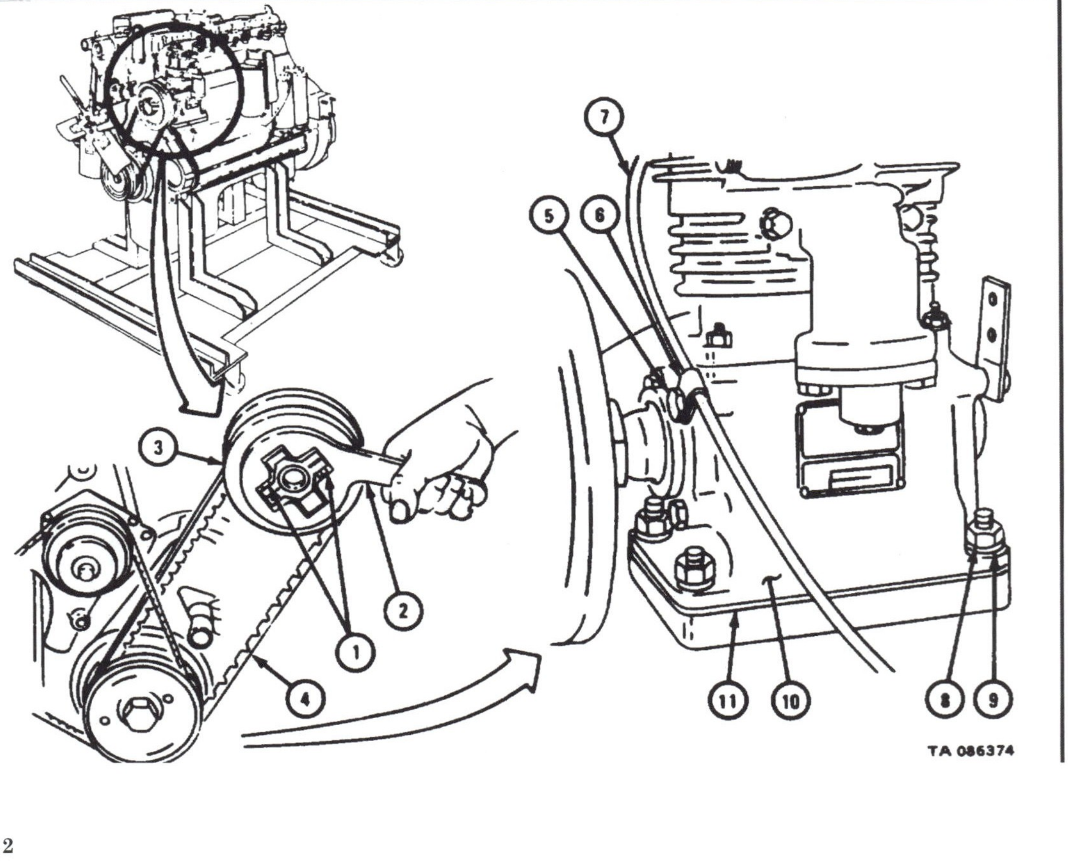

b. Engine LDS-465-2.

FRAME 1

1. Put two capscrews (1) into holes in rear face of air compressor pulley rear flange (2). Tighten capscrews and spread pulley front flange (3) apart from pulley rear flange.

GO TO FRAME 2

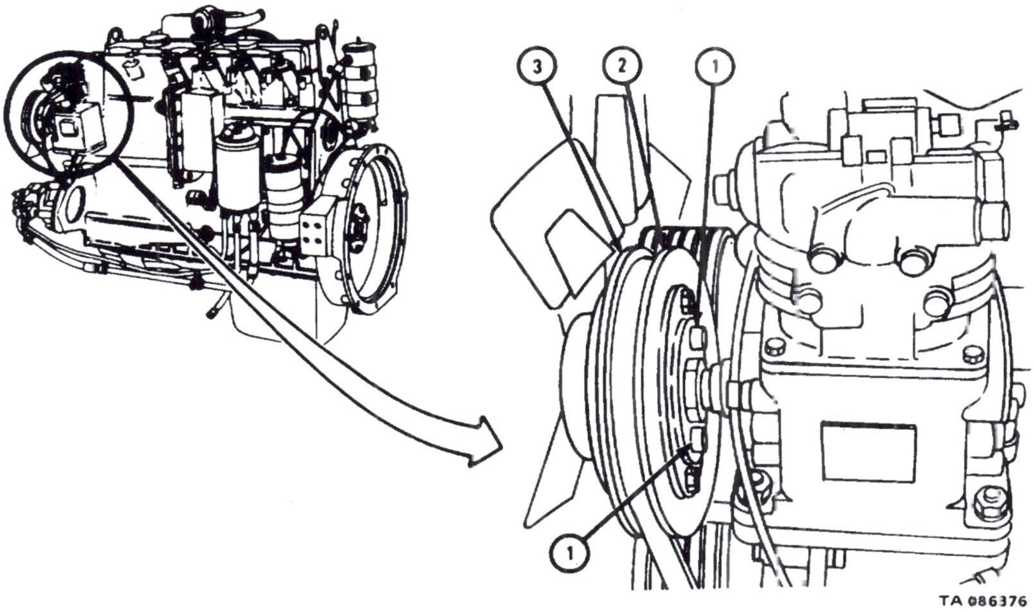

FRAME 2

1. Hold hub (1) of rear pulley flange (2). Loosen locknut (3) inside spring retainer (4). About three threads of locknut should still mate with threads of pulley shaft (5).
2. Put compressor pulley puller (6) into middle of outer pulley flange (7) as shown.
3. Turn puller (6) until pulley assembly (8) breaks loose from air compressor shaft (5). Take off locknut (3).
4. Take off pulley assembly (8) with drive belt (9). Take off and tag drive belt (9).

GO TO FRAME 3

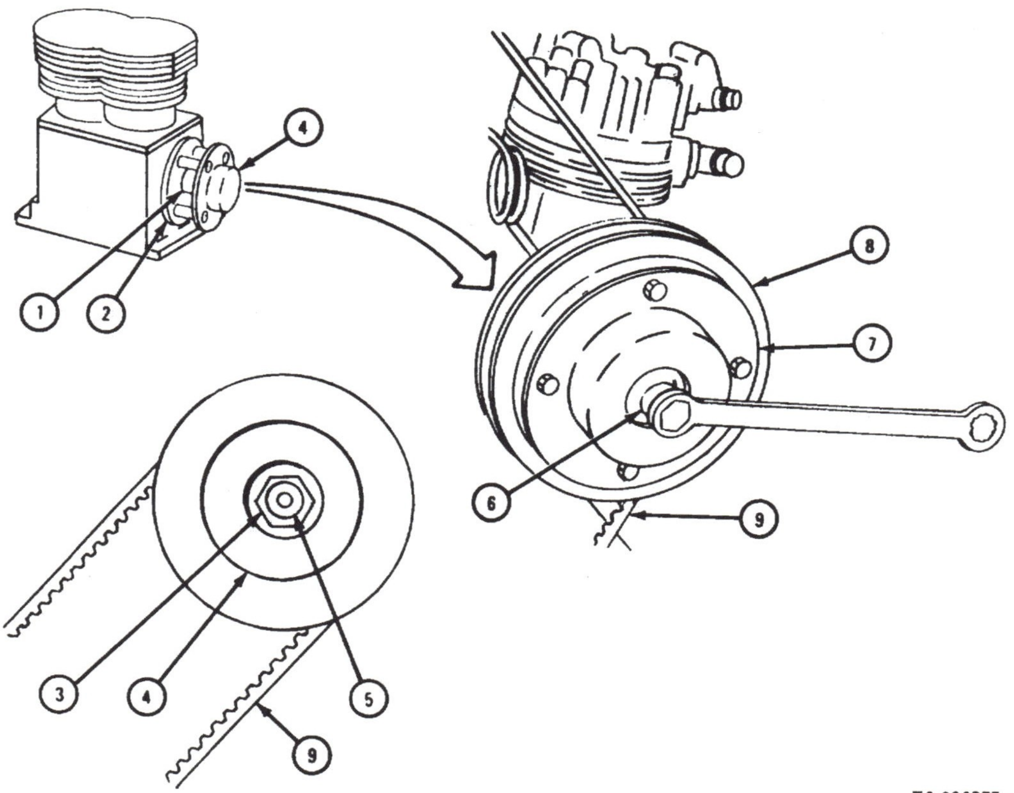

FRAME 3

1. Take out screw (1) and take off clip (2) with fuel return tube (3). Put back screw.
2. Take off four nuts (4) and lockwashers (5).
3. Lift off air compressor (6). Take off and throw away air compressor gasket (7).

END OF TASK

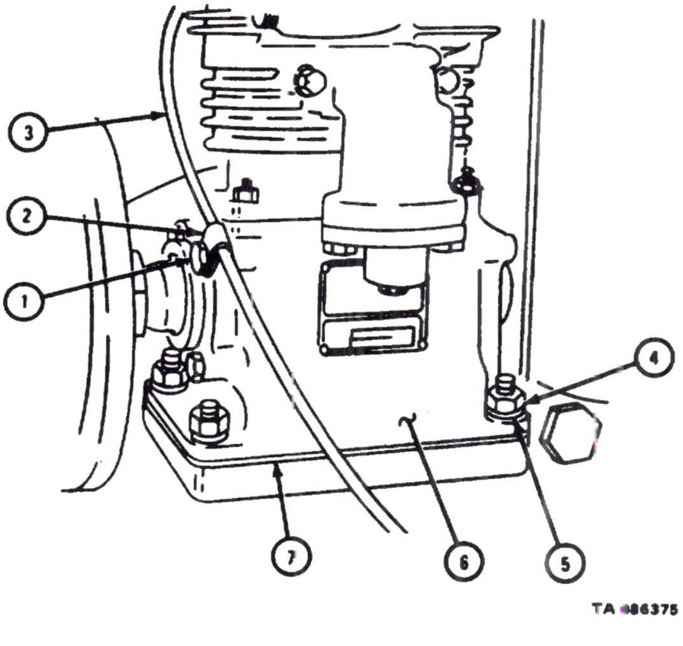

3-13. HYDRAULIC PUMP OIL RESERVOIR (ENGINE LDS-465-2).

FRAME 1

NOTE

Hydraulic pump reservoir is drained when hydraulic pump is taken off. If hydraulic pump has not been taken off, drain hydraulic oil into container when hose (2) is taken off. Empty container in approved disposal area.

Tag all hoses so they will be put back in the same place.

1. Loosen hose clamp (1). Pull off hydraulic pump oil return hose (2).
2. Loosen hose clamp (3) and pull off bypass hose (4).
3. Unscrew nut (5) and take off control valve hose (6).
4. Hold up oil reservoir (7) and take off four screws (8) and flat washers (9).
5. Take off oil reservoir (7).

END OF TASK

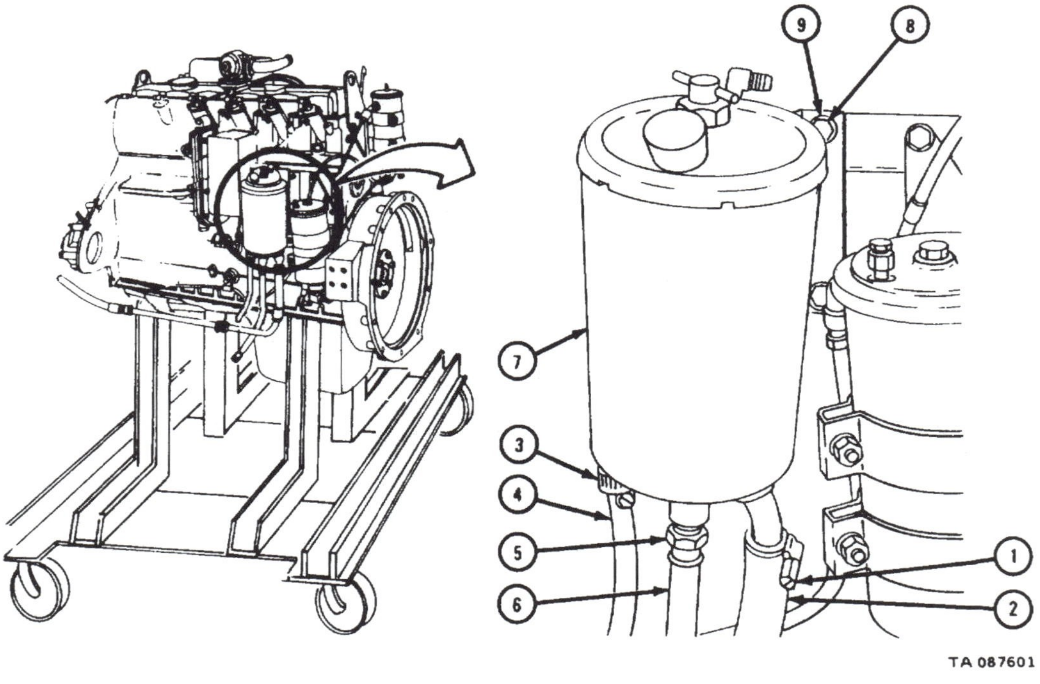

3-14. FUEL FILTER ASSEMBLY, FUEL TUBES, AND OIL HOSE.

Preliminary Procedure. Remove fuel injection pump assembly. Refer to para 3-10.

b. Engines LD-465-1, LD-465-1C, and LDT-465-1C.

FRAME 1

NOTE

Tag all tubes so that they will be put back in the same place.

This task is shown for engines LD-465-1C and LDT-465-1C. This task is the same for engine LD-465-1.

1. Unscrew two coupling nuts (1). Take off fuel return tube (2). Take off fuel filter inlet tube (3).
2. Unscrew coupling nut (4). Take off fuel filter-to-fuel density compensator tube (5).
3. Take off three nuts and lockwashers (6) and take off fuel filter assembly (7).
4. Spread open two clamps (8) and take off two drain tubes (9).

GO TO FRAME 2

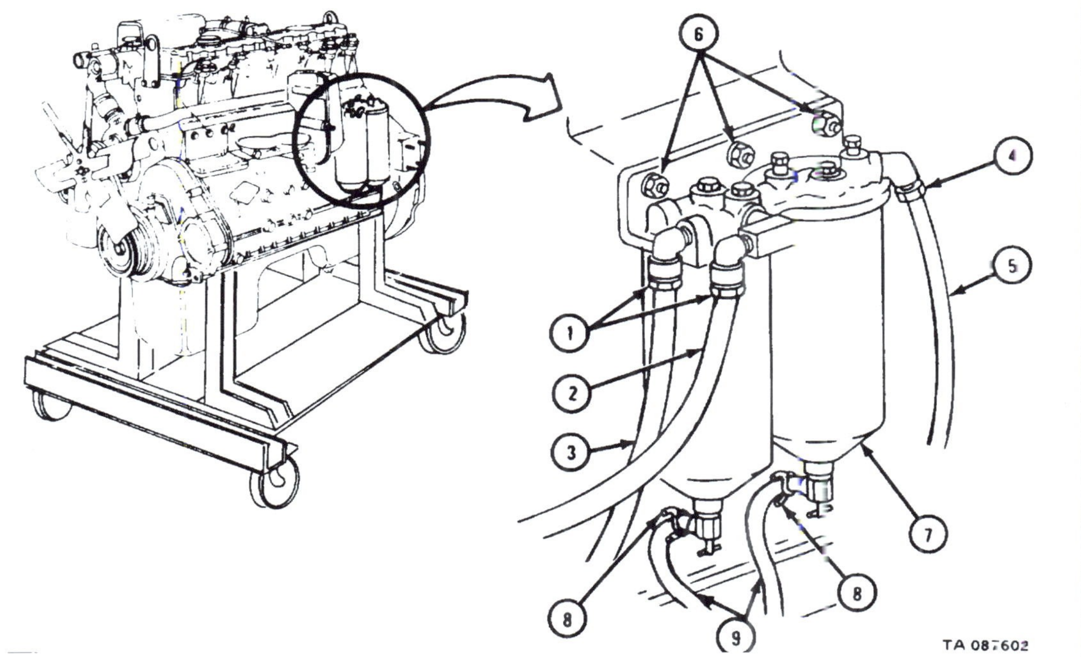

FRAME 2

1. Unscrew nut (1) and takeoff fuel injection pump pressure oil hose (2).
2. Take out pipe elbow (3).

END OF TASK

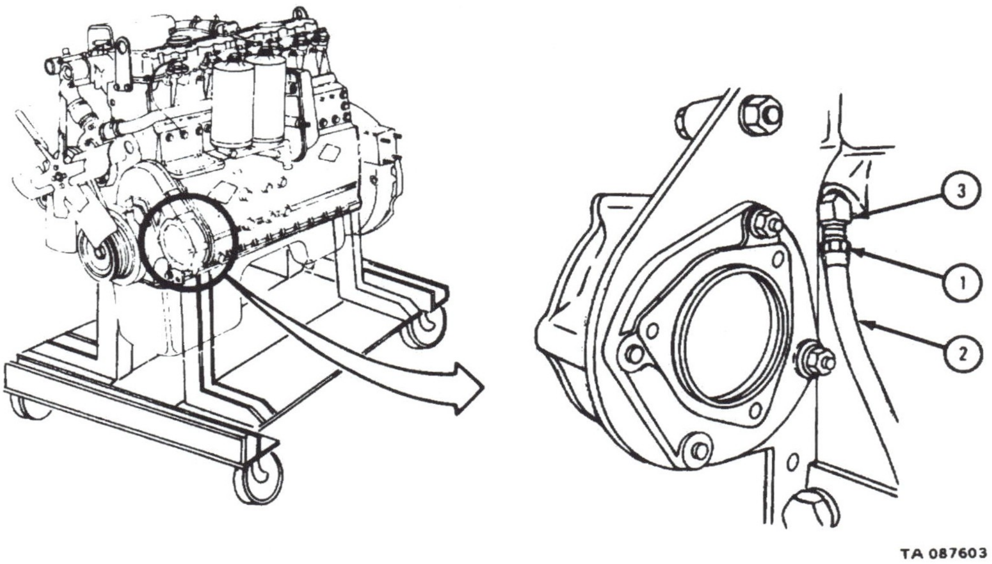

c. Engine LDS-465-1A.

FRAME 1

NOTE

Tag all tubes so they will be put back in the same place.

1. Unscrew nut (1) and take off fuel inlet tube (2).
2. Unscrew nut (3) and take off fuel return tube (4).
3. Unscrew nut (5) and take off fuel filter-to-fuel density compensator tube (6).
4. Take out two screws and lockwashers (7). Take out fuel inlet tube (2) with clamp (8) and fuel return tube (4) with clamp (9).

GO TO FRAME 2

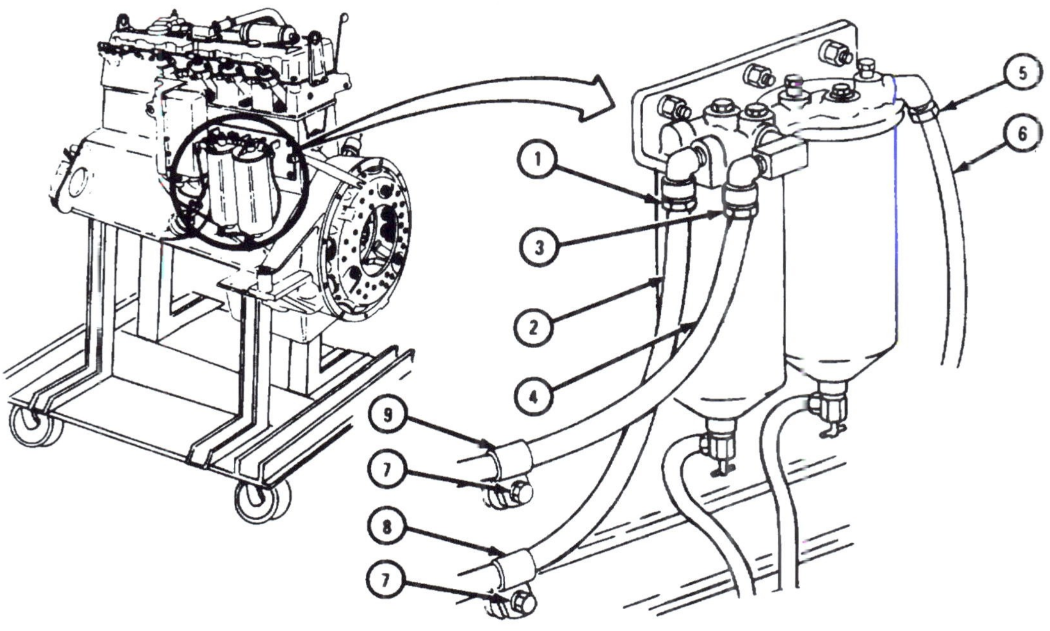

FRAME 2

NOTE

Tag all tubes so they will be put back in the same place.

1. Unscrew nut (1) and take off drain tube (2).
2. Unscrew nut (3) and take off drain tube (4).

GO TO FRAME 3

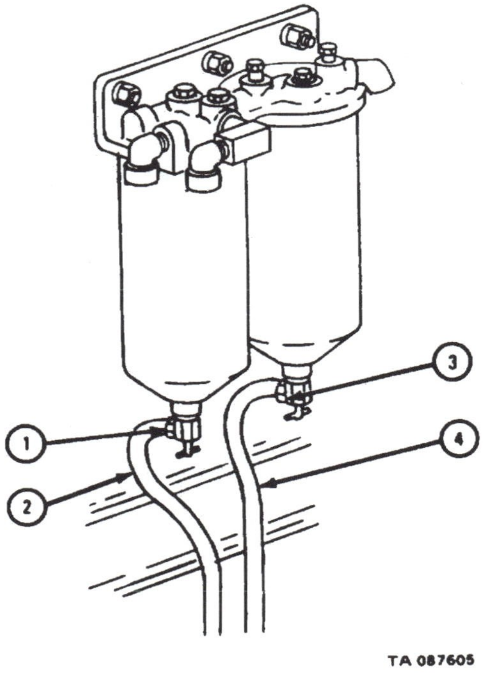

FRAME 3

1. Takeoff three nuts and lockwsshers (1) and takeoff fuel filter assembly (2) .

GO TO FRAME 4

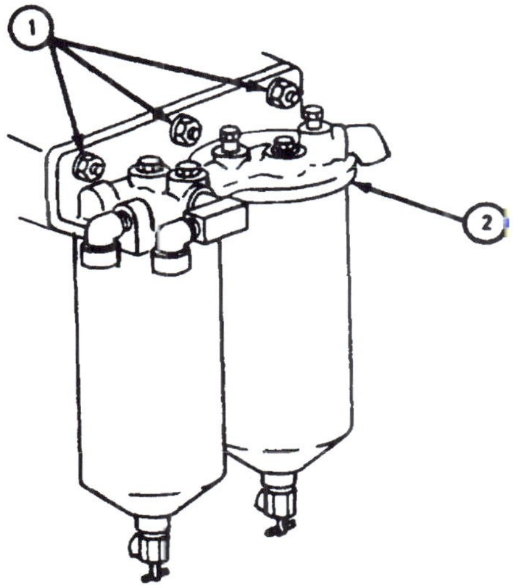

FRAME 4

1. Unscrew nut (1) and takeoff fuel injection pump Pressure oil hose (2).
2. Unscrew and take out pipe elbow (3).

END OF TASK

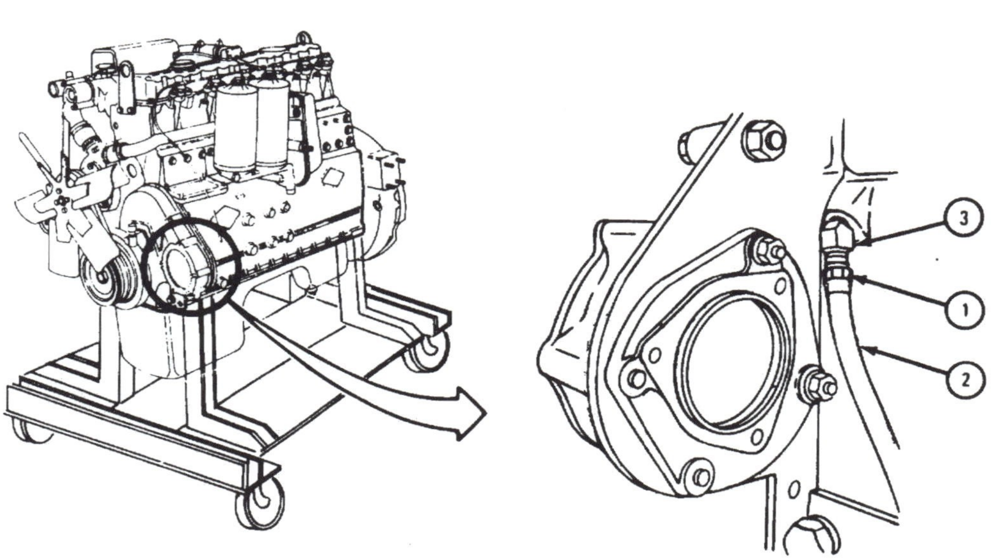

d. Engine LDS-465-1.

FRAME 1

NOTE

Tag all hoses so that they are put back in the same place.

1. Unscrew nut (1) and take off fuel inlet hose (2).
2. Unscrew nut (3) and take off fuel outlet hose (4).
3. Take out screw and lockwasher (5) and take out clamp (6).
4. Unscrew and take out fuel inlet elbow (7) and fuel outlet elbow (8).

GO TO FRAME 2

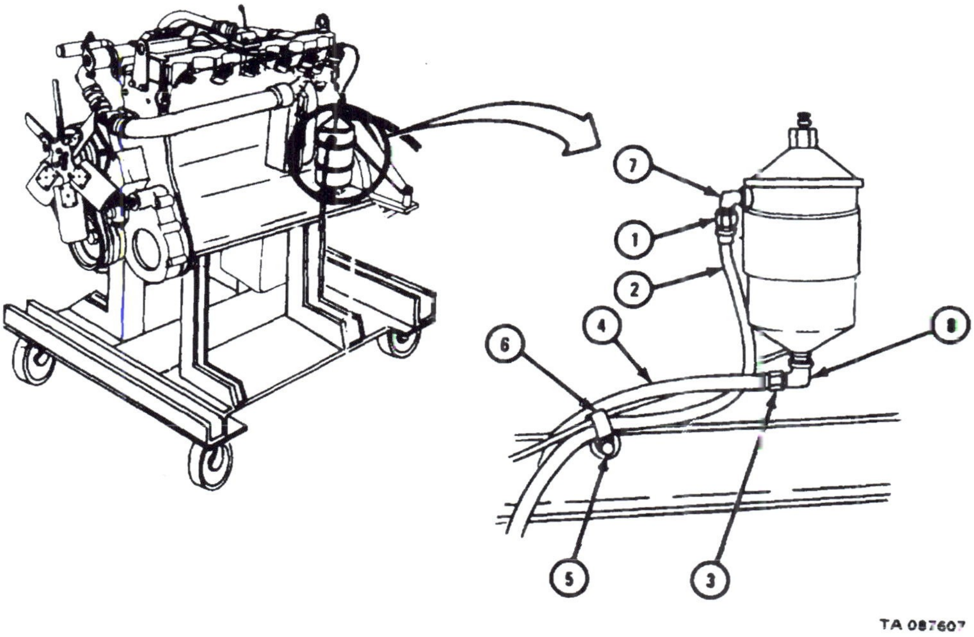

FRAME 2

1. Take out two screws and flat washers (1) and take out fuel filter assembly (2) .

GO TO FRAME 3

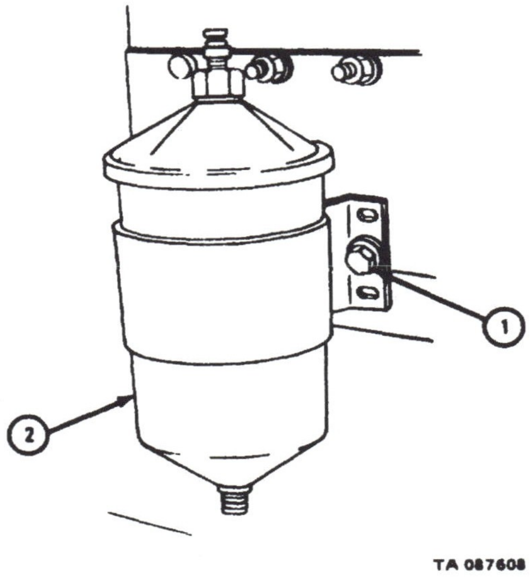

FRAME 3

NOTE

Tag all hoses so that they are put back in the same place.

1. Unscrew nut (1) and take out pressure oil hose (2).
2. Unscrew and take out elbow (3).

END OF TASK

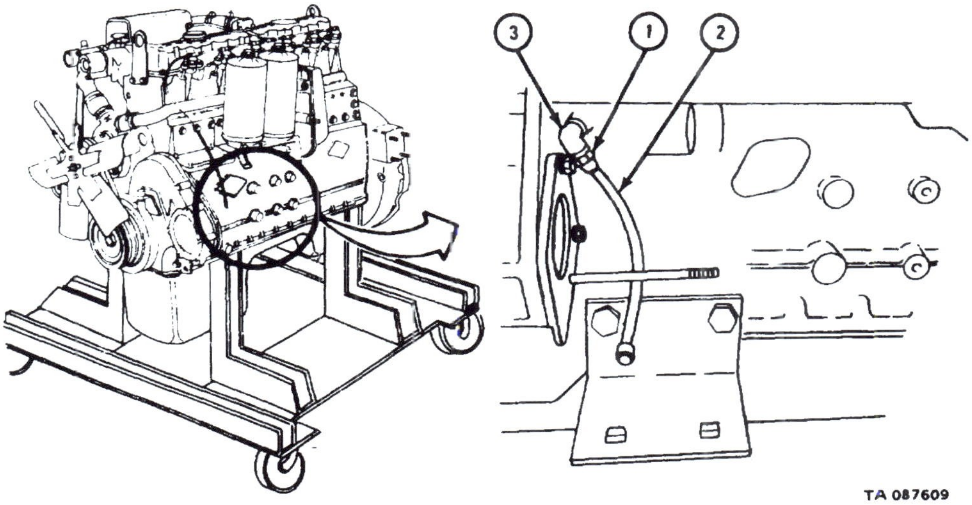

e. Engine LDS-465-2 .

FRAME 1

NOTE

Tag all tubes so that they are put back in the same place.

1. Unscrew nut (1) and take off secondary fuel filter inlet tube (2).
2. Take out screw and lockwasher (3) and take out secondary fuel filter inlet tube (2) with clamp (4) .
3. Unscrew nut (5) and take off secondary fuel filter-to-final fuel filter tube (6).
4. Unscrew nut (7) and take off other end of secondary fuel filter-to-final fuel filter tube (6) .
5. Take out screw and washer (8) and take out secondary fuel filter-to-final fuel filter tube (6) with clamp (9).
6. Pull off fuel drain tube (10).

GO TO FRAME 2

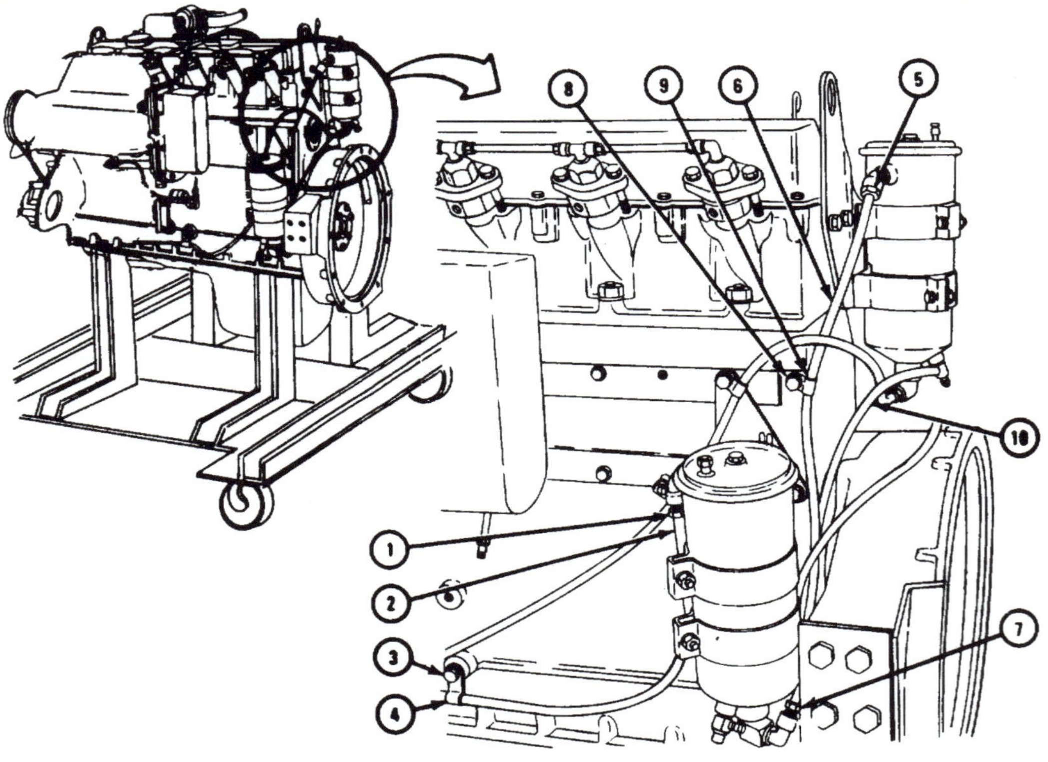

FRAME 2

NOTE

Tag all tubes so that they are put back in the same place.

1. Unscrew nut (1) and take off final fuel filter outlet tube (2).
2. Take out screw and flat washer (3) and take out final fuel filter outlet tube (2) with clamps (4 and 5).
3. Take out two screws and flat washers (6) and take out secondary fuel filter (7) with bracket (8).
4. Take out screw (9) and take out fuel drain tube (10).

GO TO FRAME 3

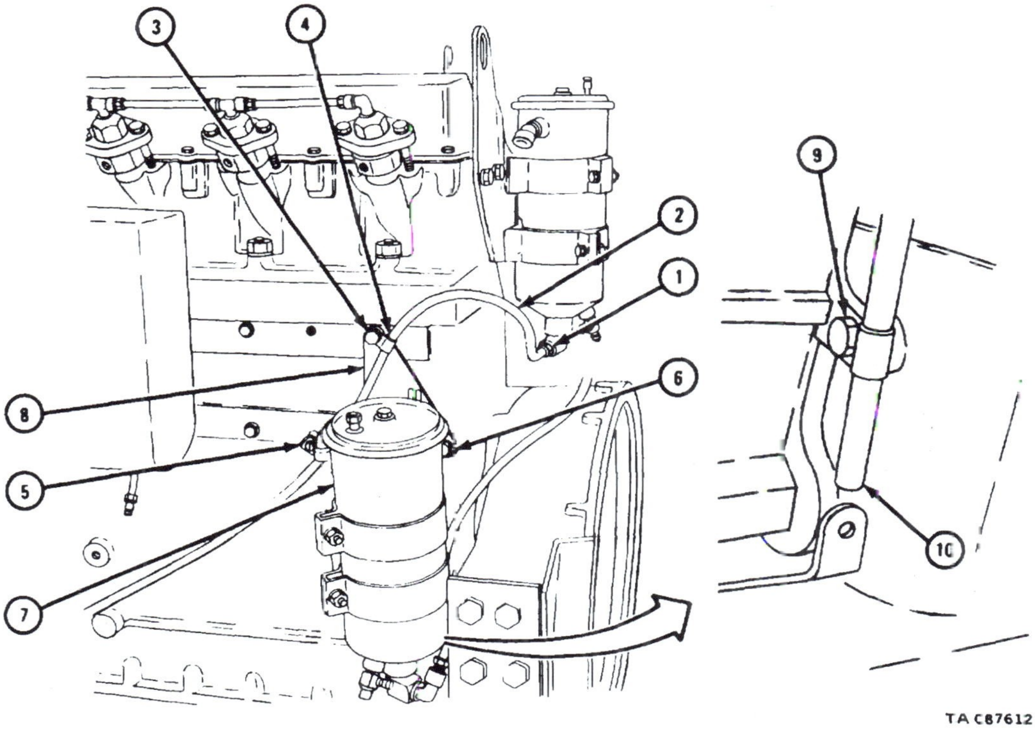

FRAME 3

NOTE

Tag pressure oi l hose (2) so that it is put back in the same place.

1. Unscrew nut (1) and take out fuel injection pump pressure oil hose (2).
2. Unscrew and take out pipe elbow (3).

END OF TASK

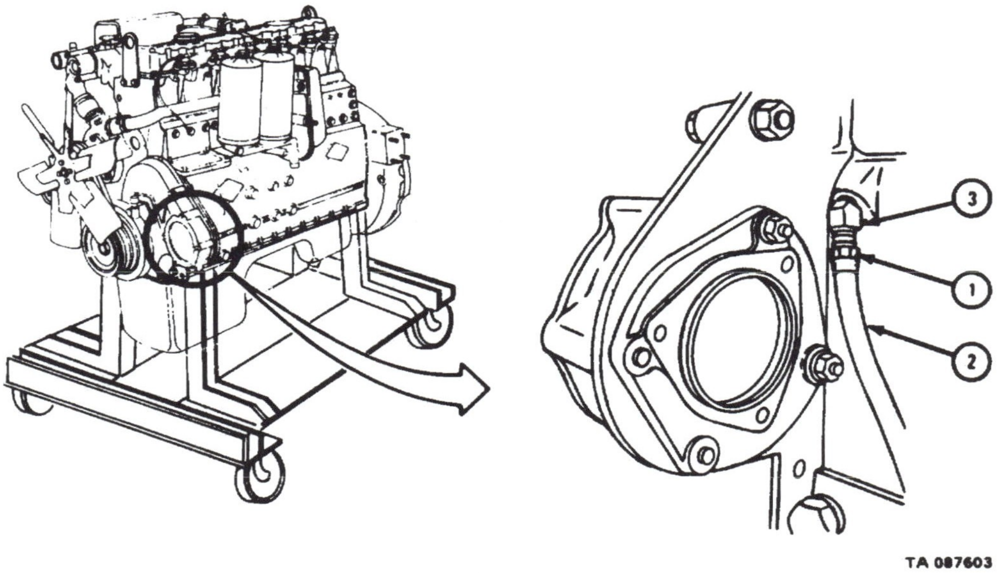

3-15. OIL COOLER AND OIL COOLER FILTER HOUSING.

a. Oil Cooler.

FRAME 1

NOTE

On some engines, oil cooler water inlet tube (1) is joined to water pump (2) with mounting flange (3). On other engines, oil cooler water inlet tube is joined to water pump with hose (4) and clamp (5).

1. For engines with mounting flange (3), take out four nuts and lockwashers (6).
2. For engines with hose (4) and clamp (5), loosen clamp.
3. Loosen hose clamp (7). Take off oil cooler water inlet tube (1).
4. For engines with mounting flange (3), take off and throw away gasket (8).

GO TO FRAME 2

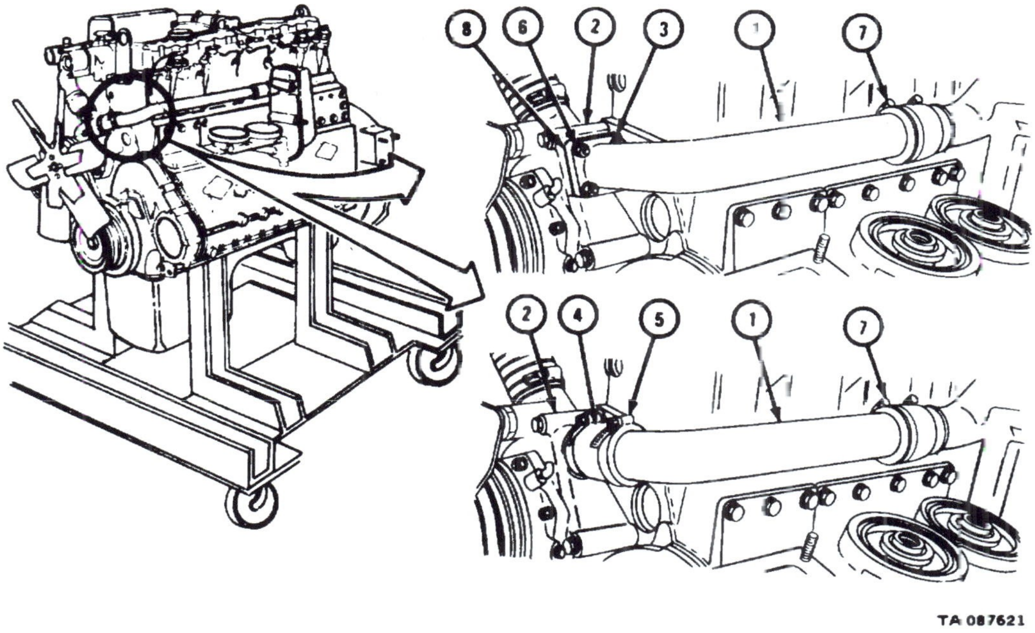

FRAME 2

NOTE

Only engines with top mounted-covered type manifold flame heater systems have tube clamp (3).

1. Take off 12 nuts (1) and lockwashers (2) and take off tube clamp (3).
2. Put container under oil cooler cover (4). Take off oil cooler cover and let any water inside cover drain into container. Take off and throw away oil cooler cover gasket (5) .
3. Take out engine oil cooler (6). Take off and throw away engine oil cooler gasket (7).

END OF TASK

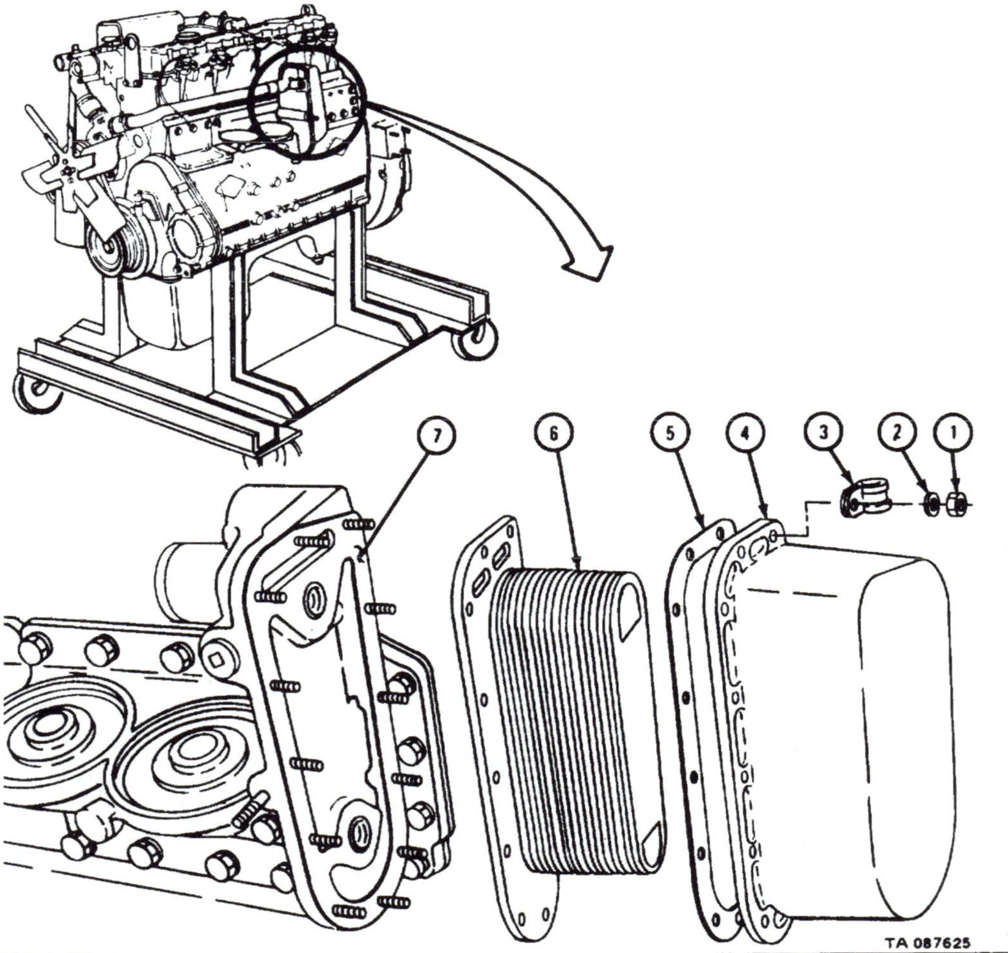

b. Oil Cooler and Filter Housing.

FRAME 1

1. Take out and throw away two preformed packing O-rings (1).
2. Take out screw and flat washer (2).
3. Take out 16 screws and flat washers (3).
4. Take out oil cooler and filter housing (4).
5. Take off and throw away oil cooler and filter housing gasket (5).

END OF TASK

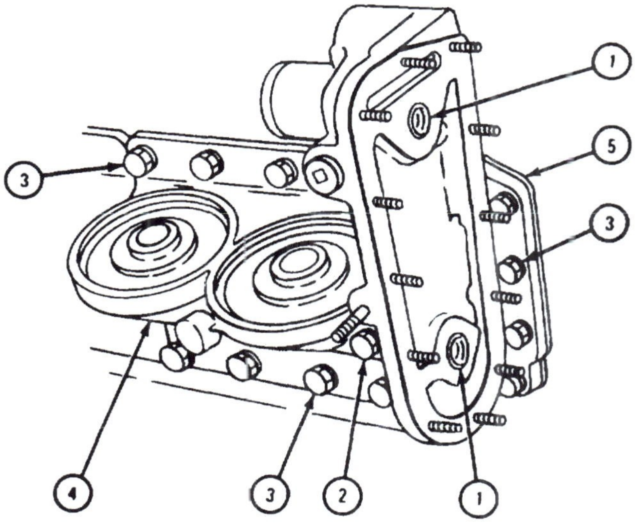

3-16. TAPPET CHAMBER COVER.

a. Engine LDS-465-1 .

FRAME 1

1. Take out two screws and flat washers (1) and take off two tube clamps (2) with flame heater fuel tubes (3 and 4).

NOTE

Some engines do not have starwasher (6) holding ground strap (5) .

2. Take off screw, flat washer, and starwasher (6) and take off ground strap (5) .

3. Take off six screws and flat washer (7).

NOTE

Two tappet chamber cover screws and lockwashers (7) were taken out during fuel filter assembly removal.

On some engines three lockwashers (8) are welded to tappet chamber cover (10) and cannot be taken off.

4. Take off three nuts (9) and lockwashers (8) and take off tappet chamber cover (10) .

GO TO FRAME 2

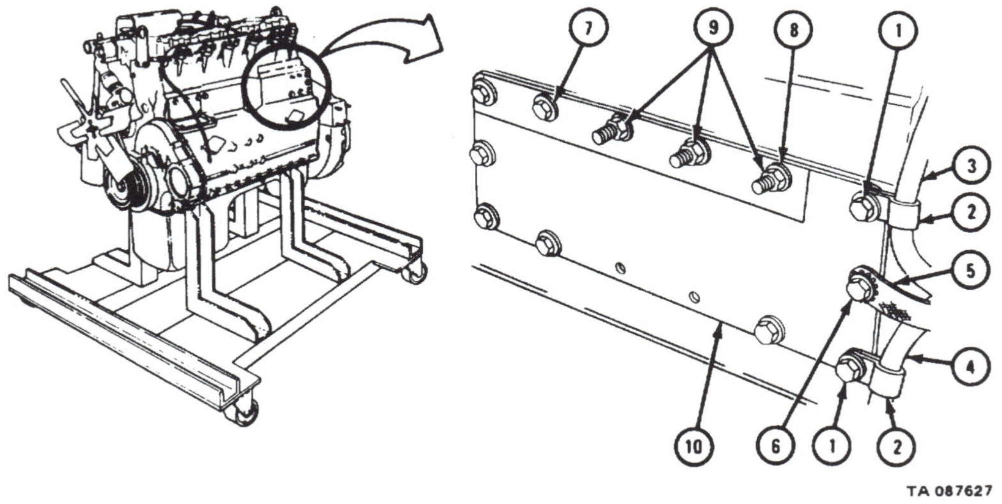

FRAME 2

1. Take off tappet chamber cover gasket (1) and throw it away.
2. If engine has crankcase water baffle (2), pull it out and throw it away.

NOTE

Crankcase water baffle (2) does not help engine performance and will not be put back.

END OF TASK

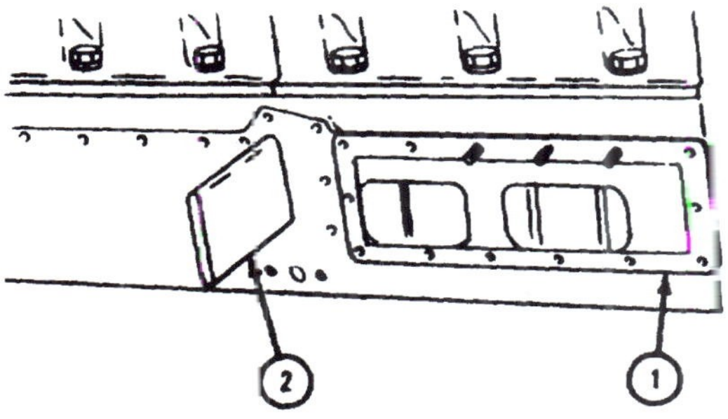

b. Engines LD-465-1, LD-465-1C, LDT-465-1C, and LDS-465-1A.

FRAME 1

1. On engines LD-465-1, LD-465-1C, and LDT-465-1C, take out screw (1), flat washer (2), and starwasher (3). Take out screws 10 screws (4) and flat washers (5) and take off tappet chamber cover (6).
2. On engine LDS-465-1A, take out screw (1), flat washer (2), starwasher (3), and ground strap (7). Take out 10 screws (4) and flat washers (5) and take off tappet chamber cover (6).

GO TO FRAME 2

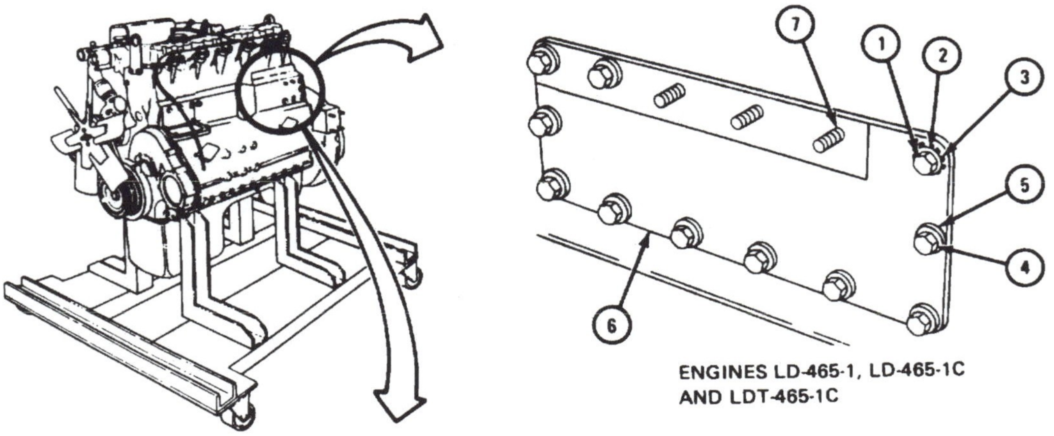

ENGINES LD-465-1, LD-465-1C AND LDT-465-1C

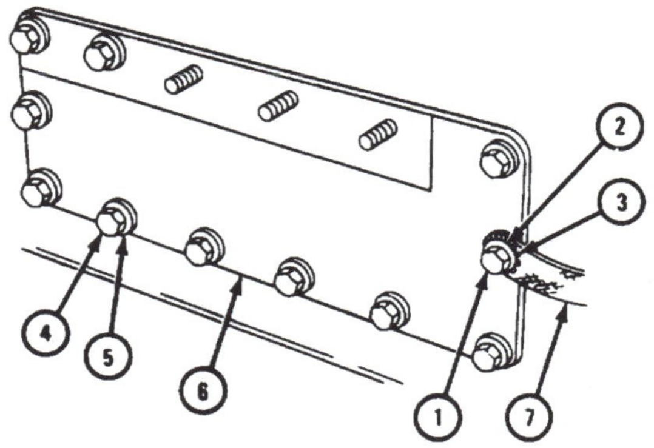

ENGINE LDS-465-1A

TA 087629

FRAME 2

1. Take off tappet chamber cover gasket (1) and throw it away.
2. If engine has crankcase water baffle (2) pull it out and throw it away.

NOTE

Crankcase water baffle (2) does not help engine performance and will not be put back.

END OF TASK

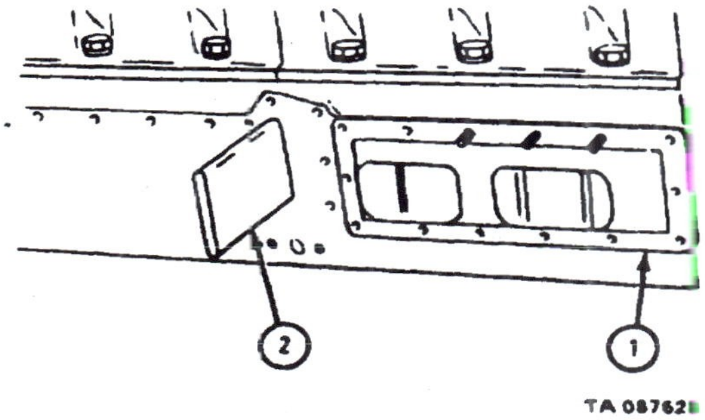

c. Engine LDS-465-2.

FRAME 1

1. Take out six screws and flat washers (1) and take off tappet chamber cover (2).

NOTE

Eight screws and flat washers were taken out of tappet chamber cover (2) during fuel filter removal and hydraulic oil reservoir removal.

GO TO FRAME 2

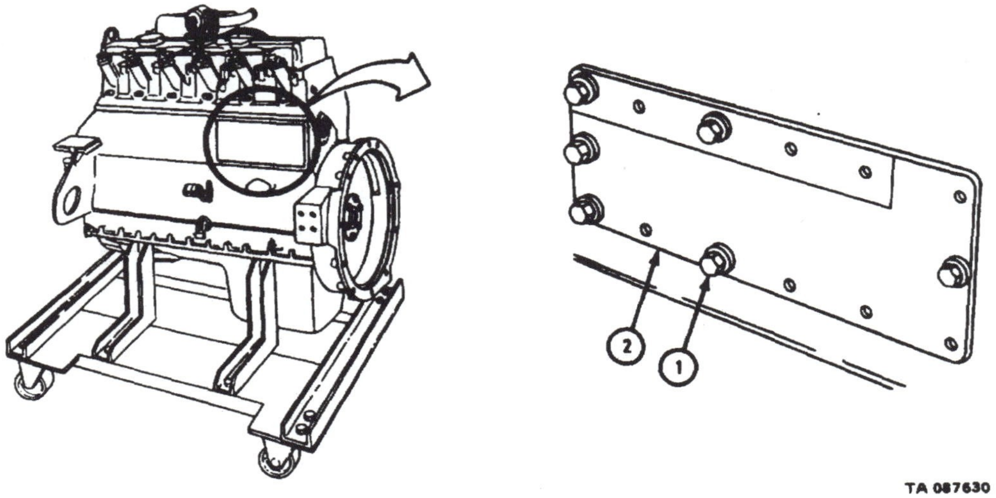

FRAME 2

1. Take off tappet chamber cover gasket (1) and threw it away.
2. If engine has crankcase water baffle (2) pull it out and throw it away.

NOTE

Crankcase water baffle (2) does not help engine. performance and will not be put back.

END OF TASK

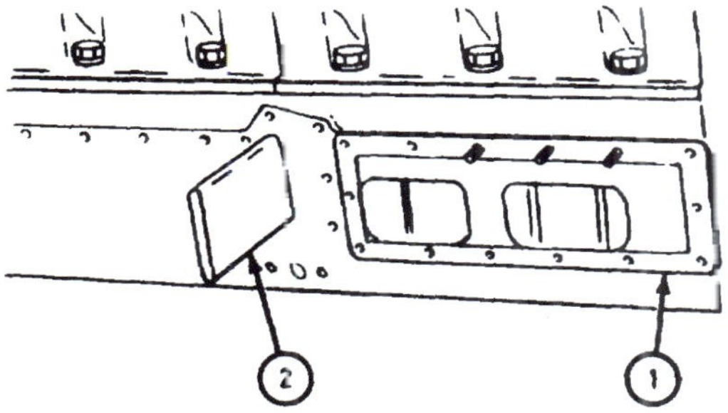

3-17. AIR COMPRESSOR SUPPORT AND FUEL INJECTION PUMP ADAPTER.

Air Compressor Support .

FRAME 1

1. Take out seven screws and flat washers (1) and take out air compressor support (2) .
2. Take off air compressor support gasket (3) and throw it away.

END OF TASK

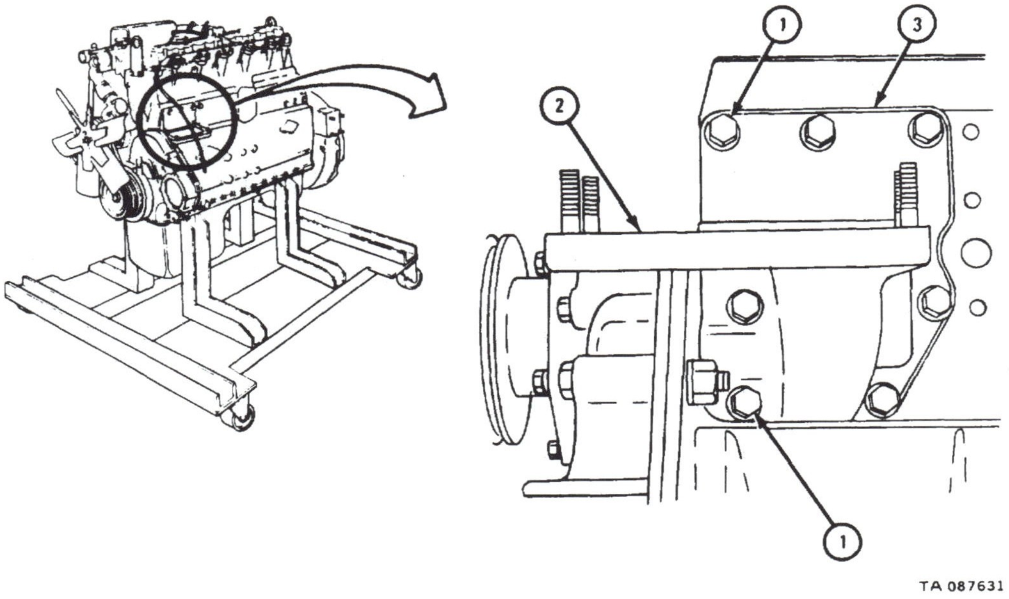

b. Fuel Injection Pump Adapter.

FRAME 1

NOTE

Some fuel injection pump adapters (1) do not have pump mounting stud (2).

1. Take off two nuts and lockwashers (3) and take out fuel injection pump adapter (1).
2. Take off and throw away fuel injection pump adapter gasket (4).

END OF TASK

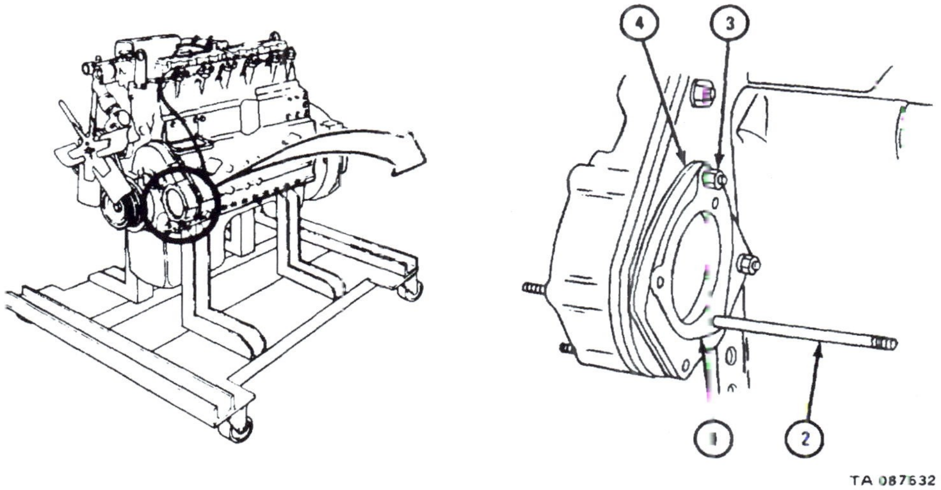

3-18. FUEL INJECTOR NOZZLE AND HOLDER ASSEMBLIES.

FRAME 1

1. Check whether routing of flame heater fuel return tube (1) is routing A or routing B .

IF FLAME HEATER FUEL RETURN TUBE (1) FOLLOWS ROUTING A, GO TO FRAME 2.
IF FLAME HEATER FUEL RETURN TUBE (1) FOLLOWS ROUTING B, GO TO FRAME 4

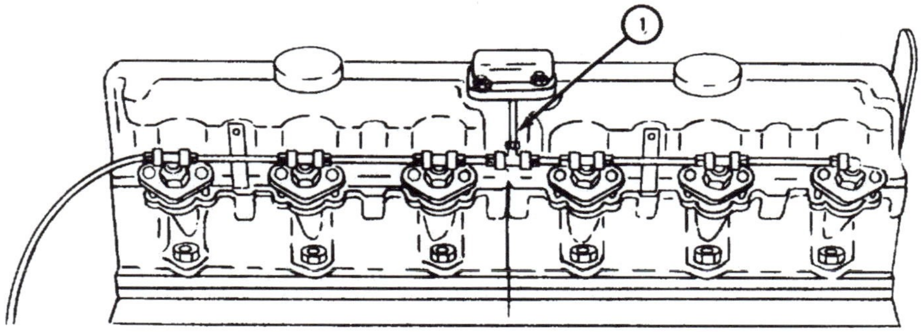

ROUTING A

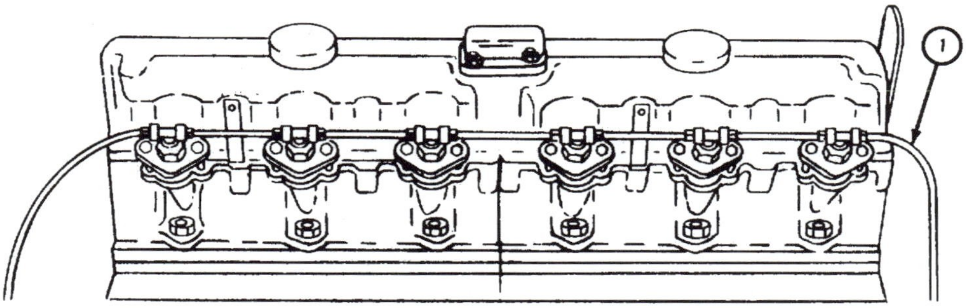

ROUTING B

TA 087614

FRAME 2

1. Unscrew nut (1) . Take off and tag flame heater fuel return tube (2).
2. Unscrew four nuts (3) . Take cut tube tee fitting (4). Takeout and tag two short fuel return tubes (5) .
3. Unscrew nut (6) . Takeout and tag fuel return-to-fuel injection pump tube (7) .

GO TO FRAME 3

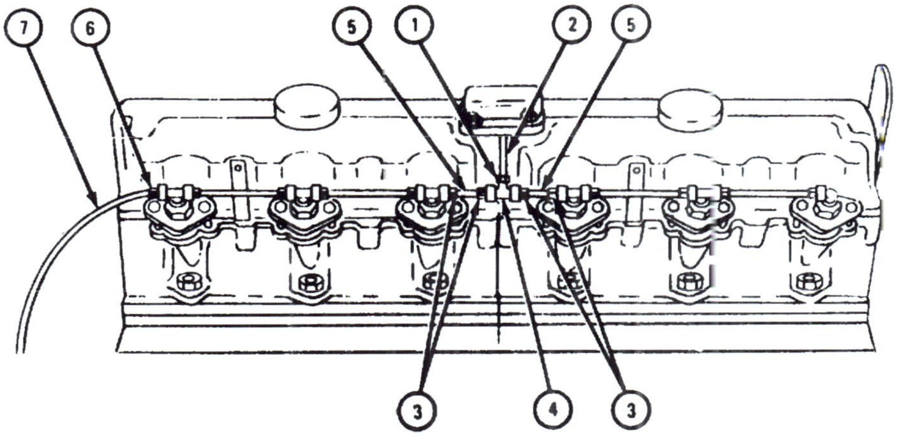

TA 087615

FRAME 3

1. Unscrew eight nuts (1) .

CAUTION

Do not turn elbow (2) or tube tee fittings (4, 6, and 8) more than 45° or fuel return tubes will break.

2. Turn elbow (2) 45° to the left and take out long fuel return tube (3). Unscrew and take out elbow (2).
3. Turn tube tee fitting (4) 45° to the left and take out long fuel return tube (5).
4. Turn tube tee fitting (6) 45° to the left and take out long fuel return tube (7).
5. Turn tube tee fitting (8) 45° to the left and take out long fuel return tube (9) .

GO TO FRAME 5

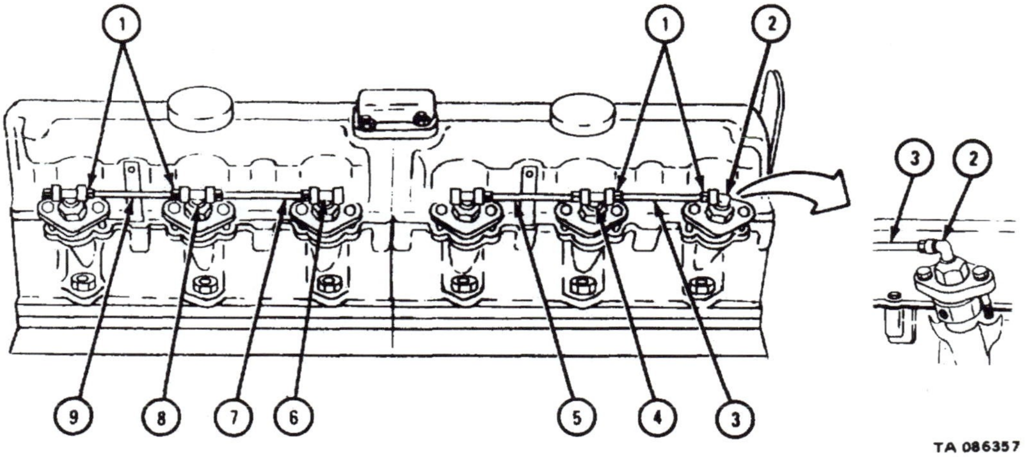

FRAME 4

1. Unscrew 12 nuts (1). Takeout and tag flame heater fuel return tube (2) and fuel return-to-fuel injection pump tube (3).

CAUTION

Do not turn tube tee fittings (4 and 6) more than 45° or fuel return tubes will break.

2. Turn tube tee fitting (4) 45° to the left as shown, and take out fuel return tube (5).
3. Do step 2 again for five tube tee fittings (6) and four fuel return tubes (7), working from rear to front of engine.

GO TO FRAME 5

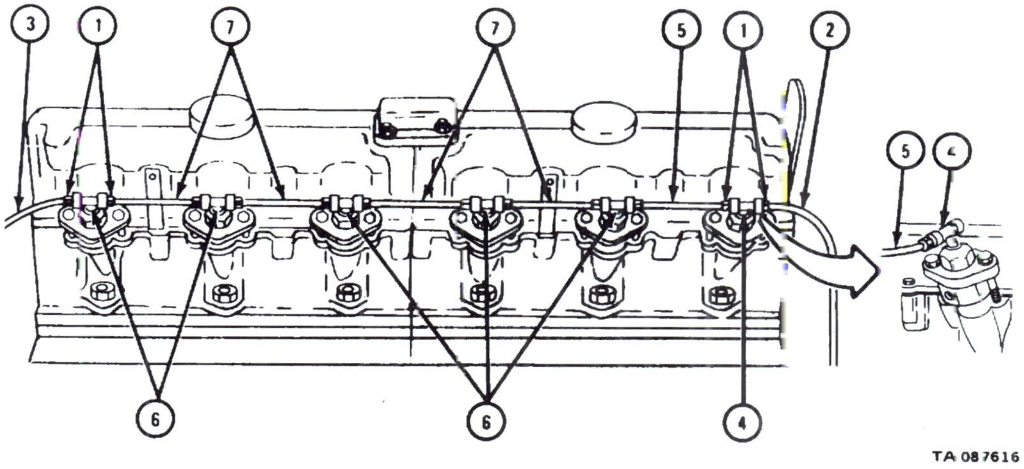

FRAME 5

1. Unscrew and take out five tube tee fittings (1).
2. Take out two screws and lockwashers (2) from each of six nozzle hold down clamps (3) .
3. Take out six nozzle hold down clamps (3).
4. Pull out six fuel injector nozzle and holder assemblies (4).

IF ANY NOZZLE AND HOLDER ASSEMBLY (4) CANNOT BE PULLED OUT,
GO TO FRAME 6.
IF ALL NOZZLE AND HOLDER ASSEMBLIES (4) CAN BE PULLED OUT,
GO TO FRAME 7

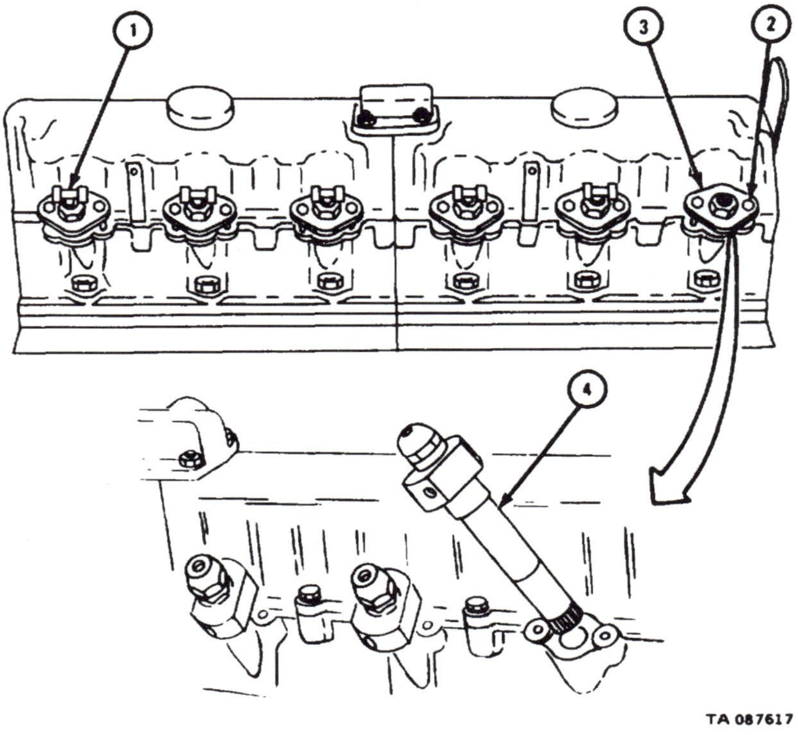

FRAME 6

NOTE

Have machine shop make nozzle and holder remover (2). See figure 3-2 .

1. Screw threaded end (1) of fabricated nozzle and holder remover (2) into end of nozzle and holder assembly (3) as shown.
2. Yank slide hammer (4) of nozzle and holder remover (2) back sharply and pull out nozzle and holder assembly (3).

GO TO FRAME 7

TA 087518

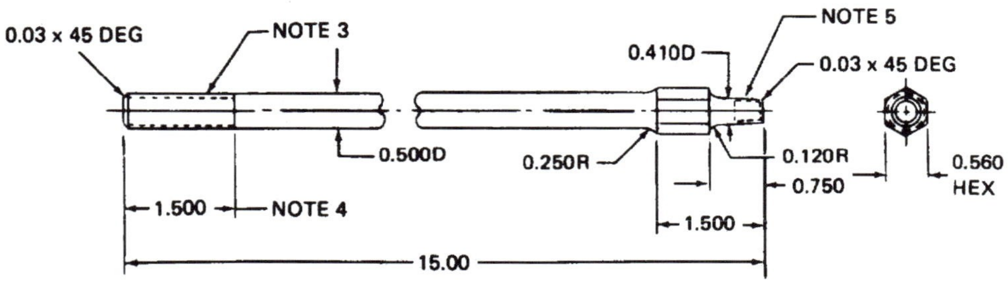

NOTES:

1. ALL DIMENSIONS GIVEN ARE IN INCHES
2. MATERIAL: CARBON STEEL SAE 1065 THRU 1095
   QUENCH AND TEMPER ROCKWELL
   "C" 30-40
3. CUT THREADS USING 1/2-20 UNF-2A DIE
4. CUT THREADS TO 1.500 IN.
5. CUT THREADS USING 1/2 -27 NPT DIE

TA 121160

Figure 3-2. Nozzle and Holder Remover Fabrication Instructions

FRAME 7

CAUTION

Plug openings (1 and 2) in fuel injector nozzle and holder assembly to keep out dirt. Dirt can clog or damage valve. inside nozzle.

1. Take off fuel injector nozzle gasket (3) and throw it away.

END OF TASK

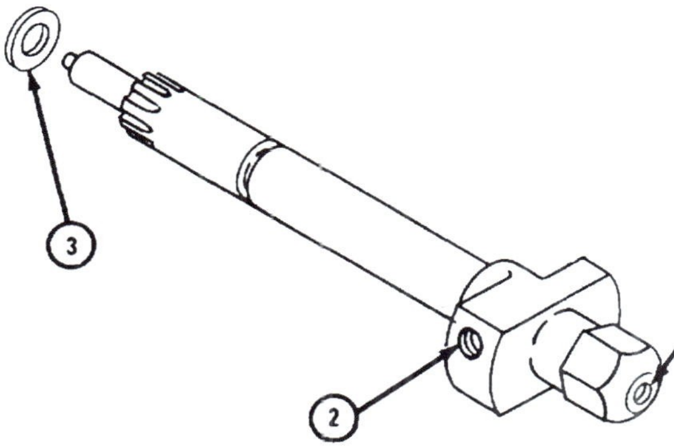

TA 087619

3-19. CRANKCASE BREATHER ADAPTER.

a. All Engines Except LDS-465-2.

FRAME 1

1. Take out four locknuts and flat washers (1). Takeoff crankcase breathe r adapter (2) with hose (3) and two clamps (4).
2. Take off and throw away two crankcase breather adapter gaskets (5).

END OF TASK

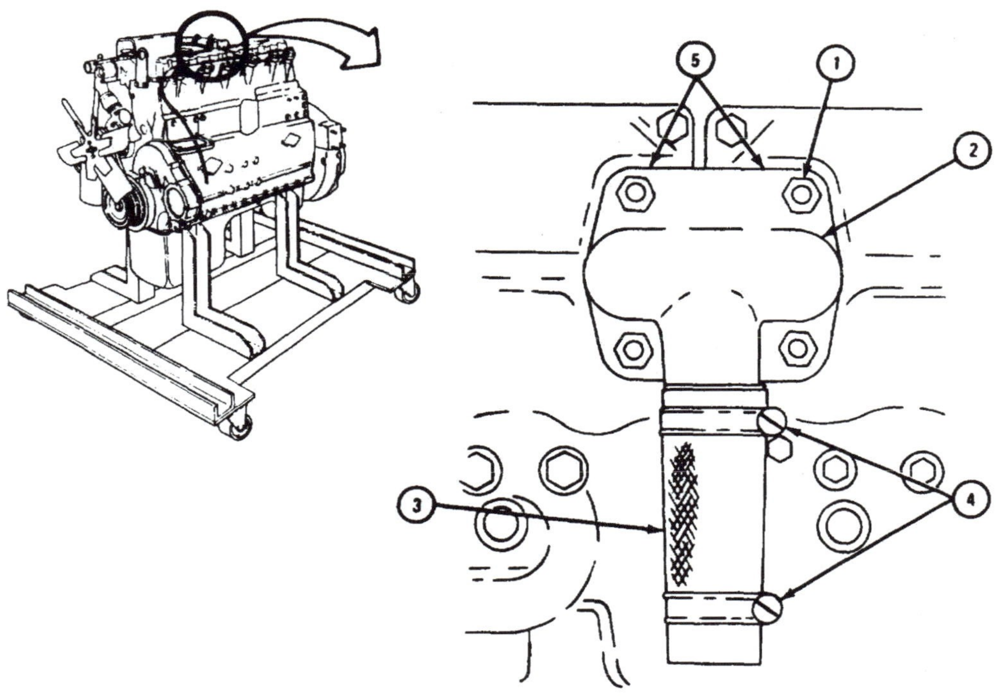

b. Engine LDS-465-2 .

FRAME 1

1. Take off four nuts and lockwashers (1) .
2. Take off crankcase breather valve (2) .
3. Take off and throw away two crankcase breather valve gaskets (3).

END OF TASK

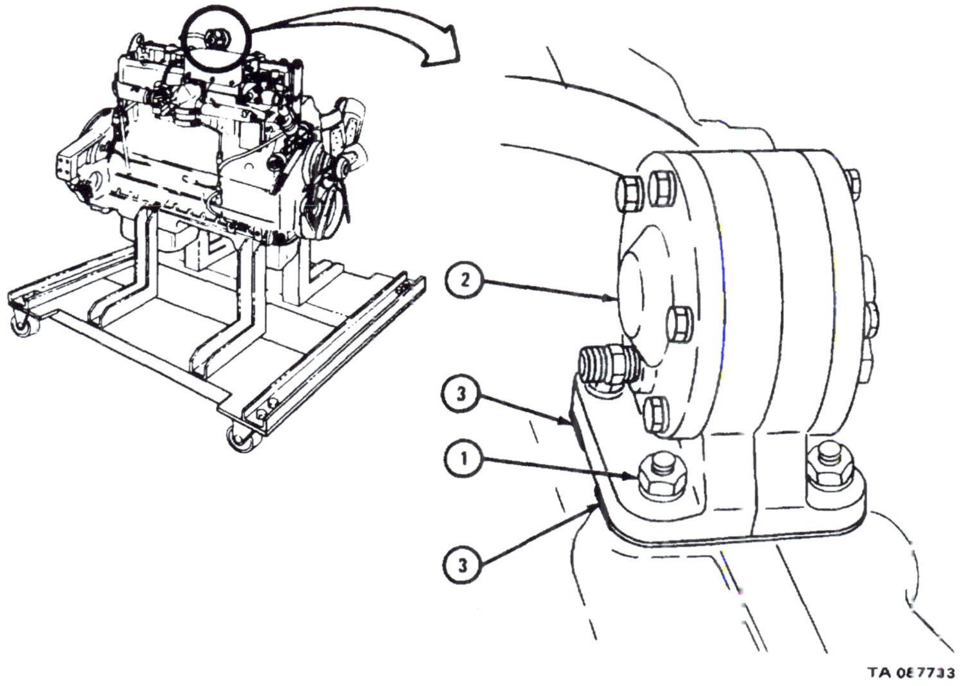

3-20. INTAKE MANIFOLD FLAME HEATER AND ELBOW.

NOTE

Intake manifold flame heater and elbow on engine LDT-465-1C were taken out when turbocharger was taken out. Refer to para 3-5.

a. Engines LD-465-1 and LD-465-1C.

FRAME 1

NOTE

This task is shown for engine LD-465-1C. This task is the same for engine LD-465-1.

1. Unscrew nut (1). Take off and tag ignition lead (2).
2. Unscrew and take out spark plug (3). Take off and throw away spark plug gasket (4).

CAUTION

Do not take out two adapters (8) using open end wrench. Adapters may be bent. Use only a tubing wrench or box wrench.

3. Take out two nuts (5). Take off and tag fuel inlet tube (6) and fuel return tube (7). Take out two adapters (8).
4. Loosen nut (9). Take out nozzle and valve assembly (10).

GO TO FRAME 2

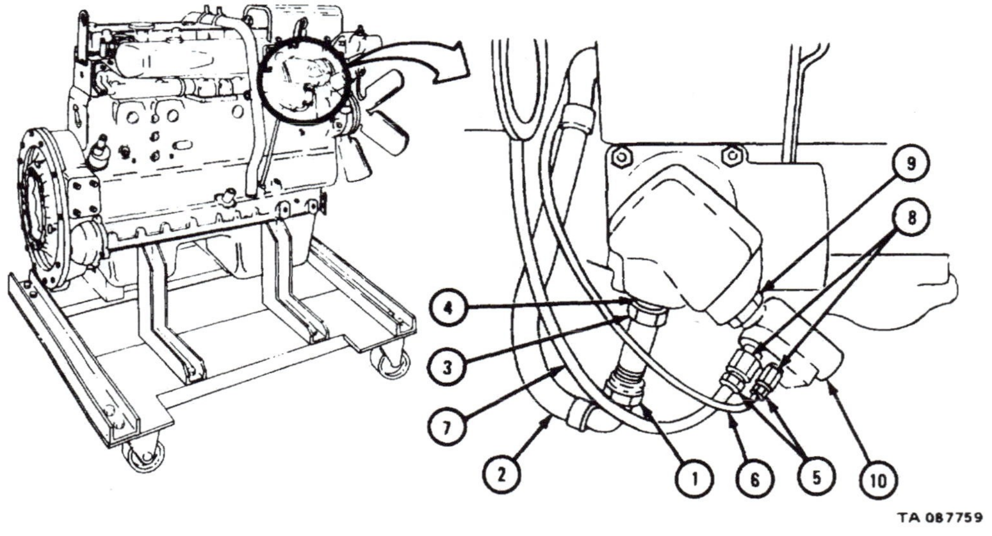

FRAME 2

1. Take off four nuts and lockwashers (1) and take out intake manifold elbow (2). Take off intake manifold elbow gasket (3) and throw it away.

END OF TASK

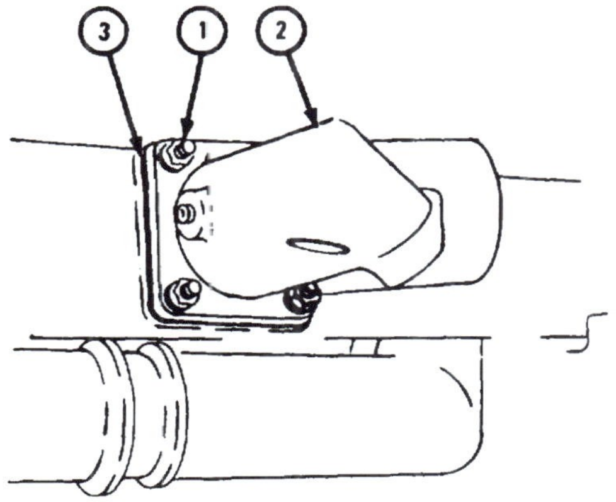

b. Engines LDS-465-1A and LDS-465-2.

FRAME 1

NOTE

This task is shown for engine LDS-465-W. This task is the same for engine LDS-465-2.

1. Unscrew nut (1). Take off and tag ignition lead.
2. Unscrew and take out spark plug (3). Take off spark plug gasket (4) and throw it away.

CAUTION

Do not take out adapters (7 and 10) using open end wrench. Adapters may be bent. Use only a tubing wrench or box wrench.

3. Unscrew nut (5). Take off and tag fuel inlet tube (6). Take out adapter (7).
4. Unscrew nut (8). Take off and tag fuel return tube (9). Take out adapter (10).

GO TO FRAME 2

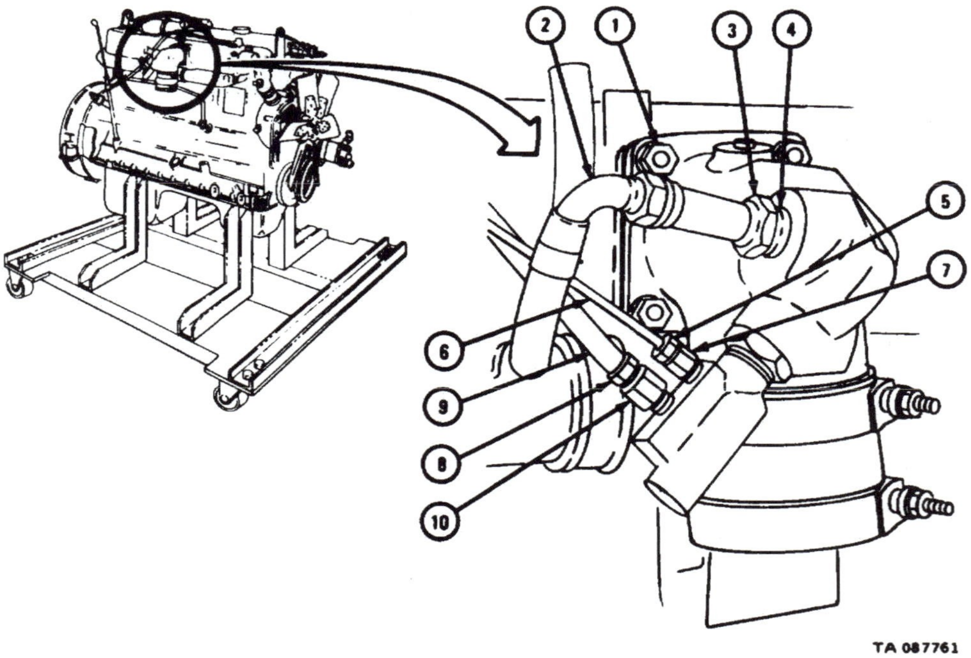

FRAME 2

1. Unscrew locknut (1) and takeout nozzle and check valve assembly (2).

2. Take off four nuts and lockwashers (3) and take out intake manifold elbow (4). Take off intake manifold elbow gasket (5) and throw it away.

END OF TASK

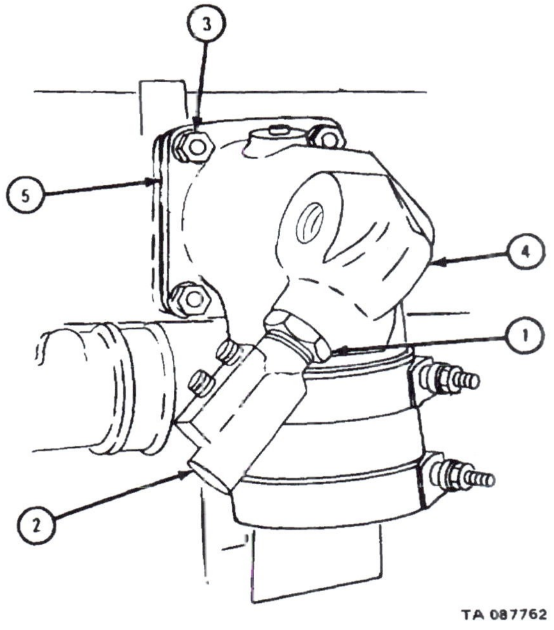

c. Engine LDS-465-1 .

FRAME 1

NOTE

AH three types of intake manifold flame heater assemblies have nozzle assembly, fuel inlet tube, fuel return tube, spark plug, and ignition lead for spark plug. These parts are in different positions on each type of flame heater assembly but are taken out the same way except where noted .

Type C flame heater assemblies may not have tube clamp (1).

1. Take out nut, lockwasher, and screw (2) and take off tube clamp (1).
2. Unscrew two nuts (3) . Take out and tag fuel inlet tube (4) and fuel return tube (5) .
3. Unscrew nut (6) and take off ignition lead (7).

NOTE

On some engines, ignition coil (8) may be mounted on top of engine instead of next to flame heater assembly as shown. If ignition coil (8) is not mounted as shown, go to frame 2.

4. Unscrew nut (9) and take ignition lead (7) off engine.

GO TO FRAME 2

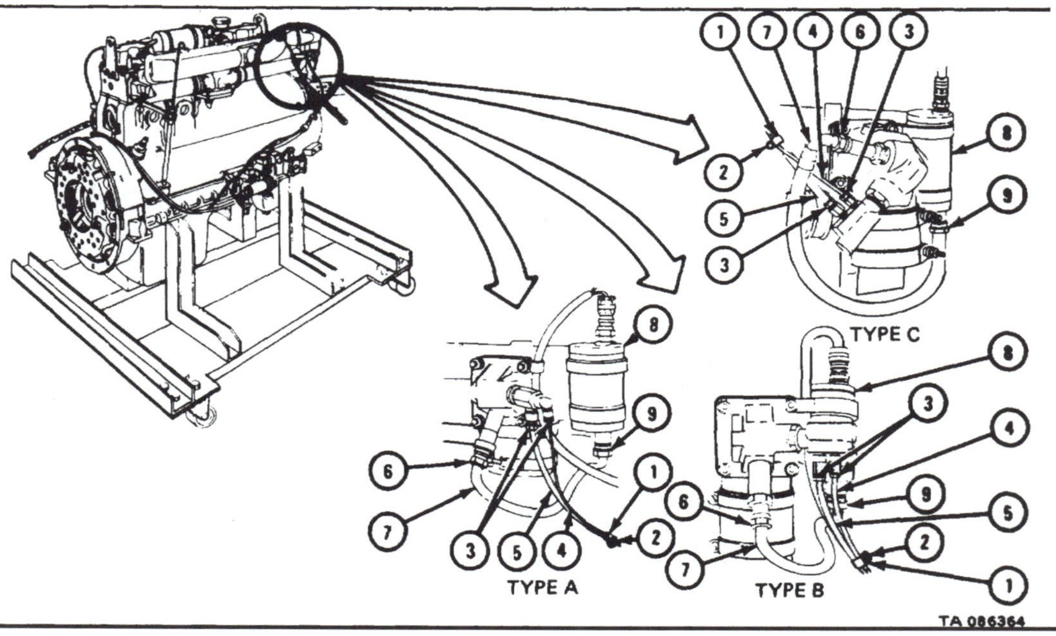

FRAME 2

1. If working on flame heater type A or B, take out four nts and lockwashers (1). Take off flame heater harness clamp (2) and flame heater assembly (3). Take off and throw away flame heater-to-intake manifold gasket (4).

NOTE

On some engines, ignition coil (5) may be mounted on top of engine instead of next to flame heater assembly as shown. If ignition coil (5) is not mounted as shown, do not do step 2.

2. Unscrew nut (6) and take out wiring harness (7).

IF WORKING ON FLAME HEATER TYPE C, GO TO FRAME 3.
IF WORKING ON FLAME HEATER TYPE A OR B, GO TO FRAME 4

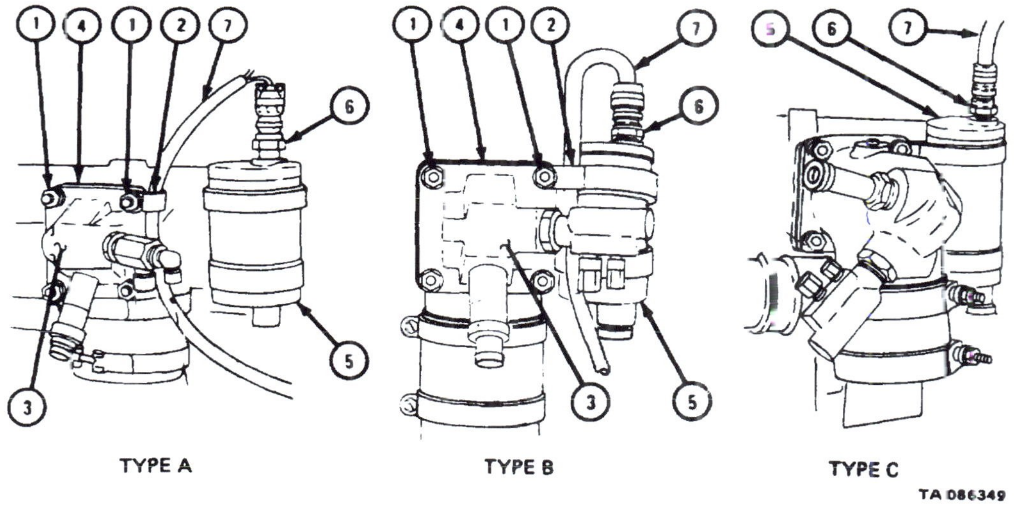

FRAME 3

1. Unscrew and takeout sparkplug (1) . Take off and throw away spark plug gasket (2) .

CAUTION

Do not take out two adapters (3) using open end wrench. Adapters may be bent. Use only tubing wrench or box wrench.

2. Unscrew and takeout two adapters (3) .

3. Loosen locknut (4) . Unscrew and take out nozzle and check valve assembly (5) .

NOTE

On some engines, ignition coil (6) may be mounted on top of engine instead of next to flame heater elbow as shown.

4. Take off four nuts and lockwashers (7) . Take off intake manifold elbow (8) and ignition coil (6) with two clamps (9). Take two clamps off ignition coil.

5. Take off and throw away intake manifold elbow gasket (10).

GO TO FRAME

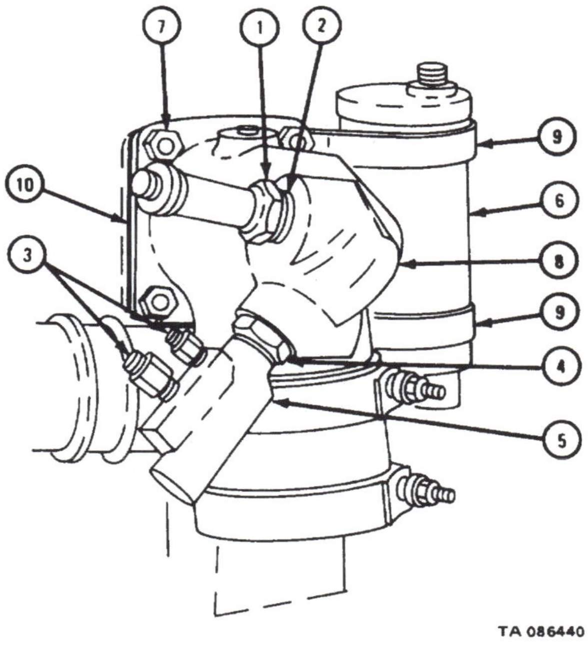

FRAME 4

1. Take off two nuts and lockwashers (1) and take off ignition coil (2) with two clamps (3) . Take two clamps off ignition coil.
2. Loosen and take off two hose clamps (4). Take off hose (5) .
3. Take off two nuts and lockwashers (6) and take off intake manifold elbow (7). Take off and throw away intake manifold elbow gasket (8).

GO TO FRAME 5

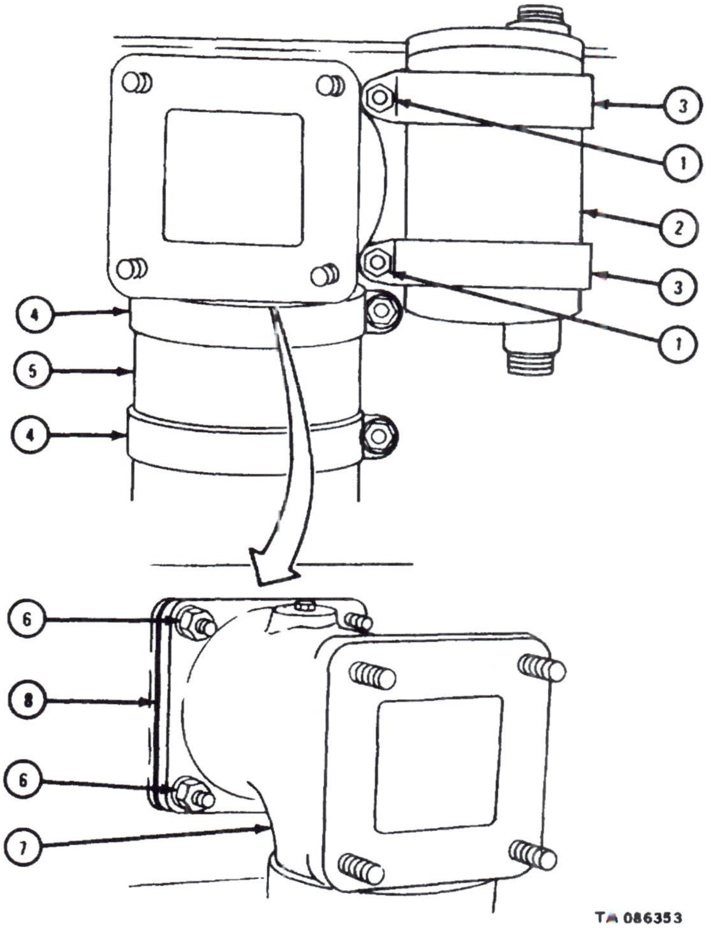

TA 086353

FRAME 5

CAUTION

Do not use open end wrench to take out pipe bushing (1) or tube adapter (2). Bushing and adapter might be bent. Use only tubing wrench or box wrench.

1. For type A flame heater assembly, unscrew and take out 90° elbow (3) and pipe bushing (1). Take out tubing adapter (2).

CAUTION

Do not use open end wrench to take out tubing adapters (4). Adapters might be bent. Use only tubing wrench or box wrench.

2. For type B flame heater assemblies, unscrew and takeout two tubing adapters (4).
3. Loosen locknut (5) and take out nozzle and check valve assembly (6).
4. Unscrew and takeout spark plug (7). Take out and throw away spark plug gasket (8).

END OF TASK

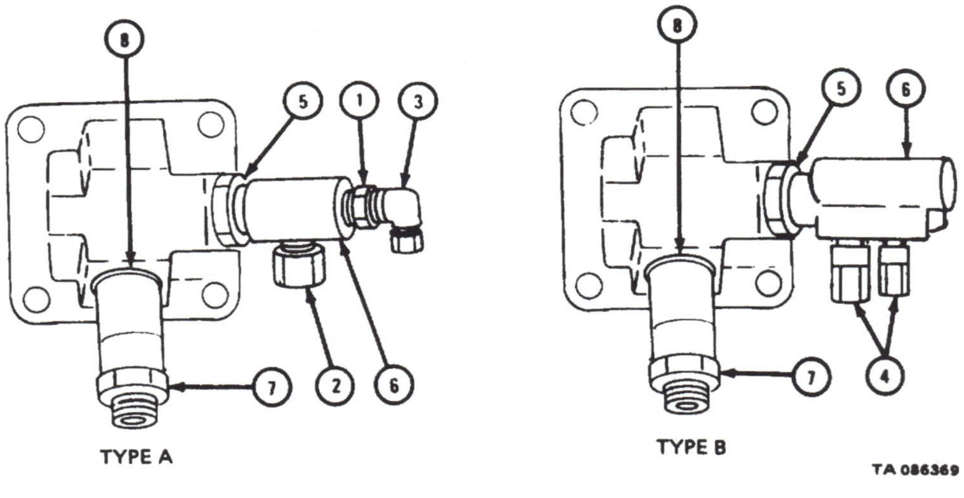

3-21. FLAME HEATER IGNITION UNIT AND FLAME HEATER FUEL PUMP.

a. Preliminary Procedure. Remove intake manifold flame heater elbow assembly. Refer to para 3-20.

b. System Identification.

FRAME 1

1. Look at top front of engine. If engine has cover (1) in position shown, engine uses top mounted-covered type flame heater system. Go to para 3-21c.
2. Look at top rear of engine. If engine has flame heater ignition unit (2) in position shown, engine uses top mounted-uncovered type flame heater system. Go to para 3-21d.
3. Look at lower right front side of engine. If engine has flame heater fuel pump assembly (3) as shown, engine uses side mounted type flame heater system. Go to para 3-21e.

END OF TASK

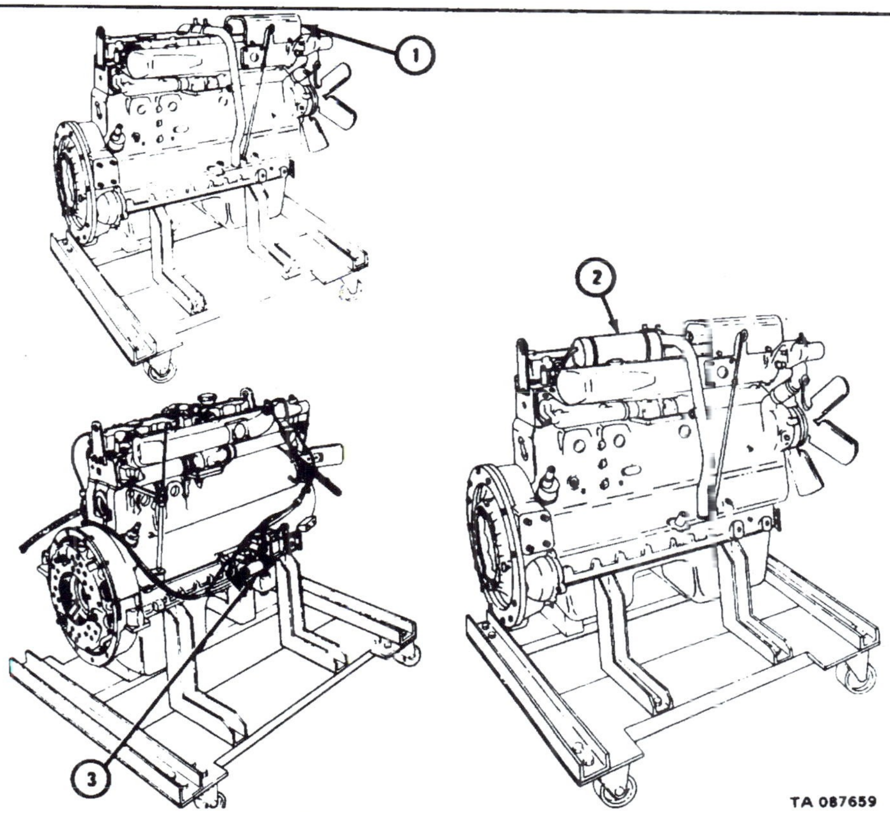

c. Top Mounted-Covered Flame Heater Ignition Unit and Fuel Pump.

FRAME 7

NOTE

Air pressurization tube (1) is used only on engine LDS-465-2.

1. Take out four screws and lockwashers (2) and take off flame heater fuel pump cover (3) .

GO TO FRAME 2

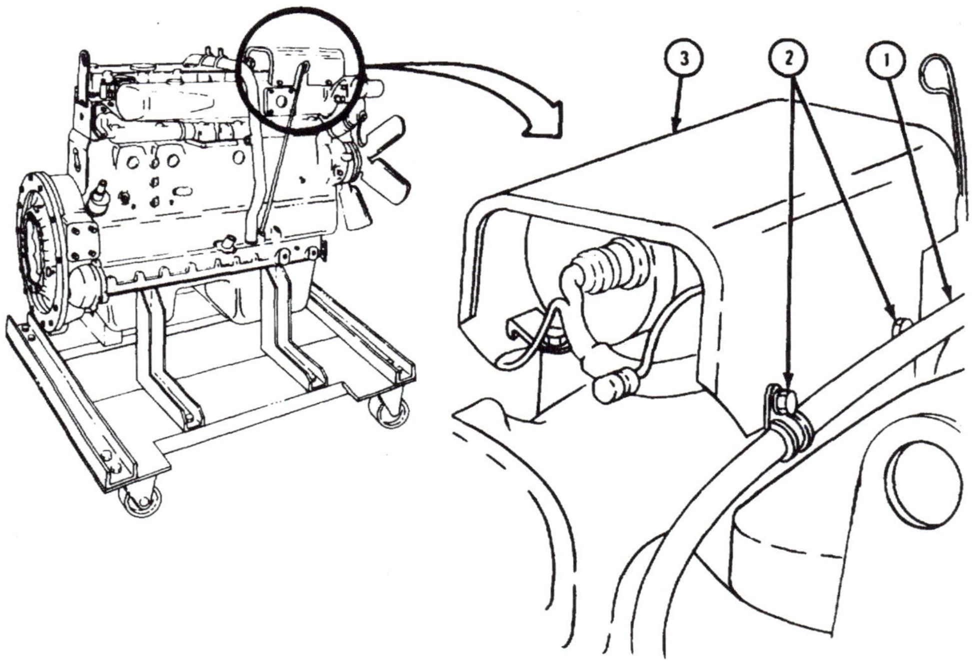

FRAME 2

NOTE

There are two types of top mounted-covered flame heater fuel pump and ignition units. Type A and type B assemblies have the same parts and are taken out in the same way.

1. Unscrew nut (1). Take out and tag ignition lead (2).
2. Unscrew nut (3). Take off and tag flame heater harness (4).
3. Take off two spring clamps (5) and take out ignition unit (6).

GO TO FRAME 3

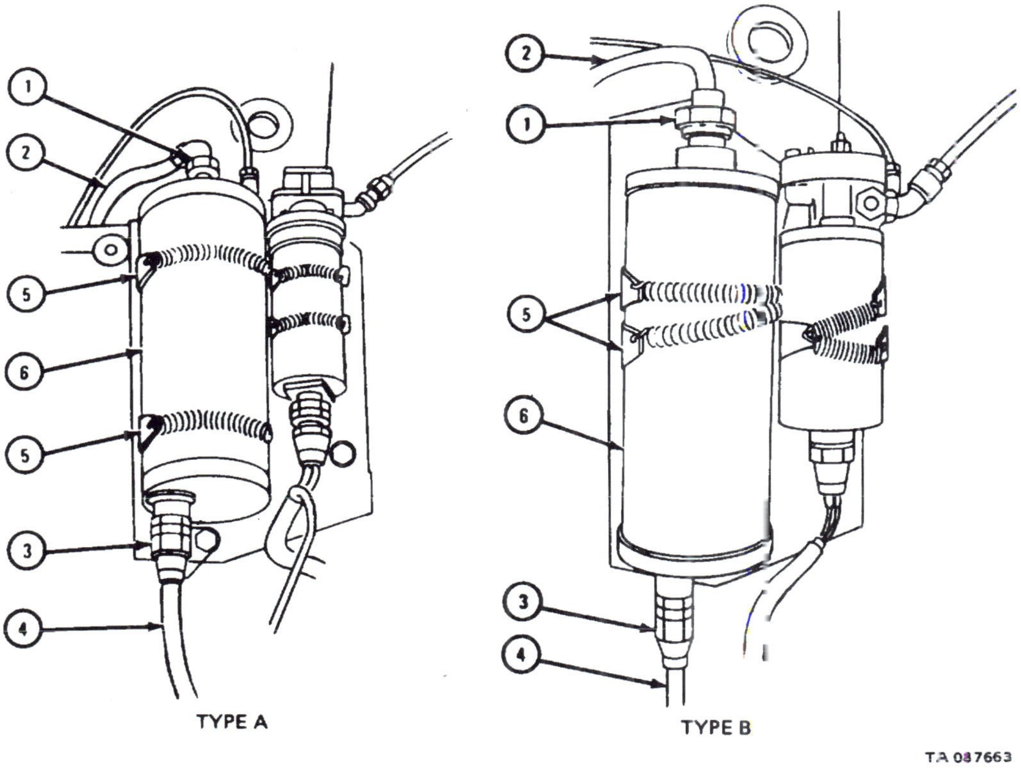

FRAME 3

NOTE

Tag all fuel and electrical lines so that they are put back in the same place.

1. Unscrew nut (1) and take off flame heater fuel filter-to flame heater fuel pump tube (2) .
2. Unscrew nut (3) and take out flame heater fuel pump-to- flame heater nozzle tube (4) .
3. Unscrew nut (5) and take off flame heater fuel pump wiring harness (6).
4. Take off two spring clamps (7) and take out fuel pump (8).
5. Unscrew and take out 90° elbow (9) and 45° elbow (10).

GO TO FRAME 4

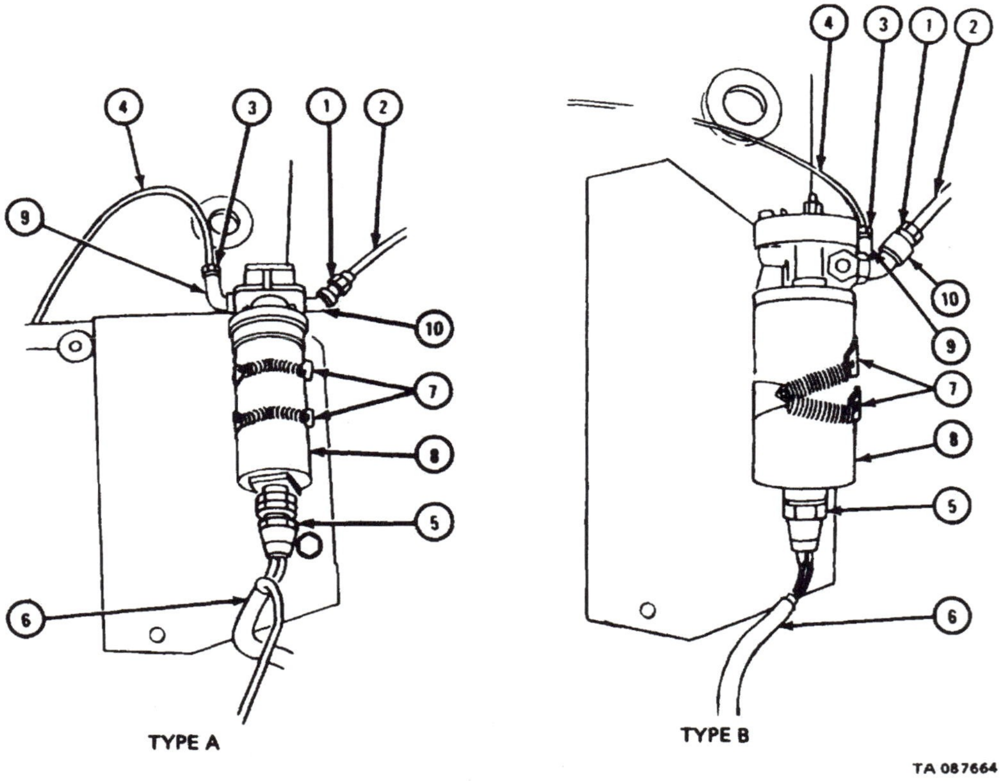

FRAME 4

1. Take out screw and lockwasher (1) . Take off flame heater harness ground lead (2) and take out flame heater harness (3).
2. Take out two screws and lockwashers (4) and take out flame heater support bracket (5) .

GO TO FRAME 5

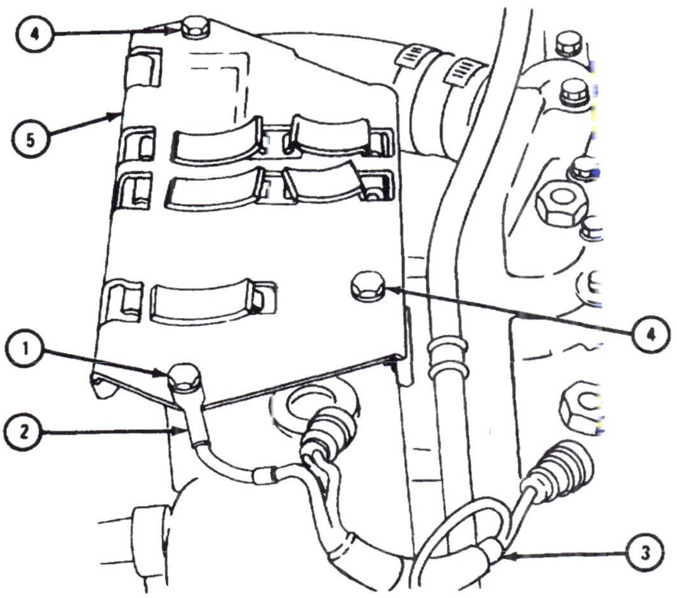

FRAME 5

NOTE

Tag all fuel lines so that they are put back in the same place.

1. Hold adapter (1), unscrew nut (2) and take out fuel filter-to-fuel pump tube (3) .
2. Unscrew nut (4) and take out fuel supply pump-to- flame heater fuel tube (5).
3. Take out screw and lockwasher (6) and take out flame heater fuel filter (7).

END OF TASK

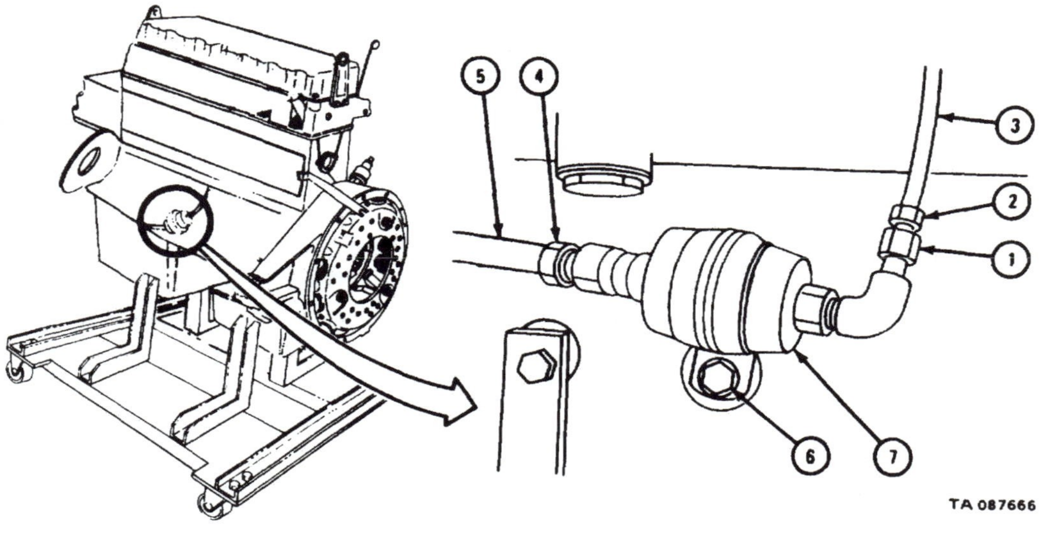

d. Top Mounted-Uncovered Flame Heater Ignition Unit and Fuel Pump.

FRAME 1

NOTE

Tag all leads so that they are put back in the same place.

1. Unscrew nut (1) and take out flame heater ignition lead (2).
2. Unscrew nut (3) and take off flame heater harness (4).
3. Take out screw and lockwasher (5) and take off flame heater harness ground lead (6) .
4. Take out screw and lockwasher (7) and take out ignition unit (8).

NOTE

Some engines with high profile water manifolds have 0.620-inch thick spacers under two clamps (9). These should be taken out with ignition unit (8).

5. Take off two clamps (9).

GO TO FRAME 2

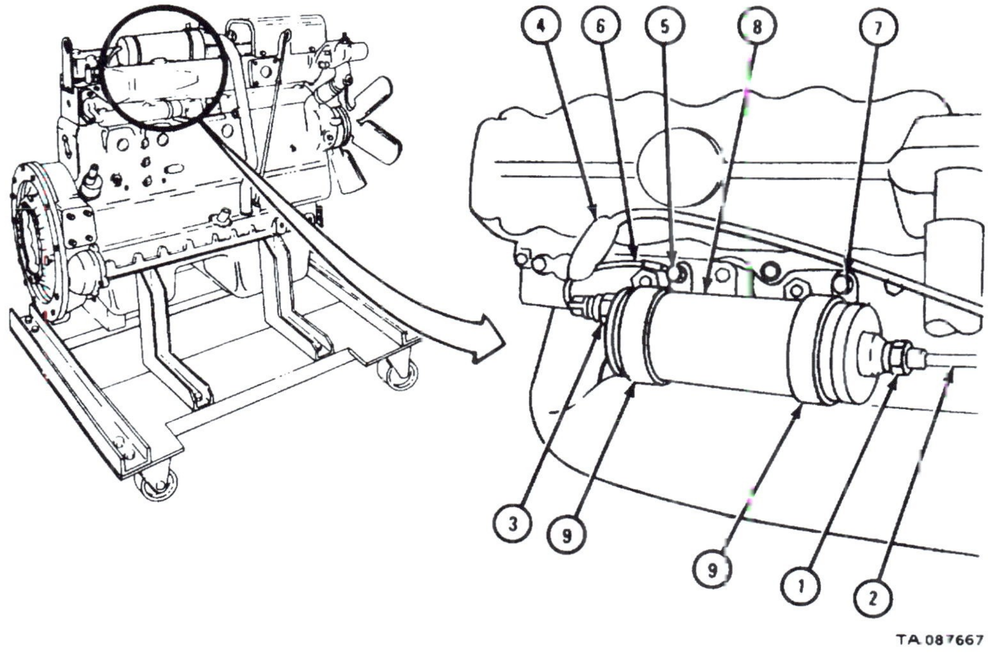

FRAME 2

NOTE

Tag all lines so that they are put back in the same place.

1. Unscrew nut (1) and take out fuel supply pump-to-flame heater filter tube (2).
2. Take out flame heater fuel filter (3) from elbow (4).
3. Take out elbow (4).
4. Take out pipe fitting (5).

GO TO FRAME 3

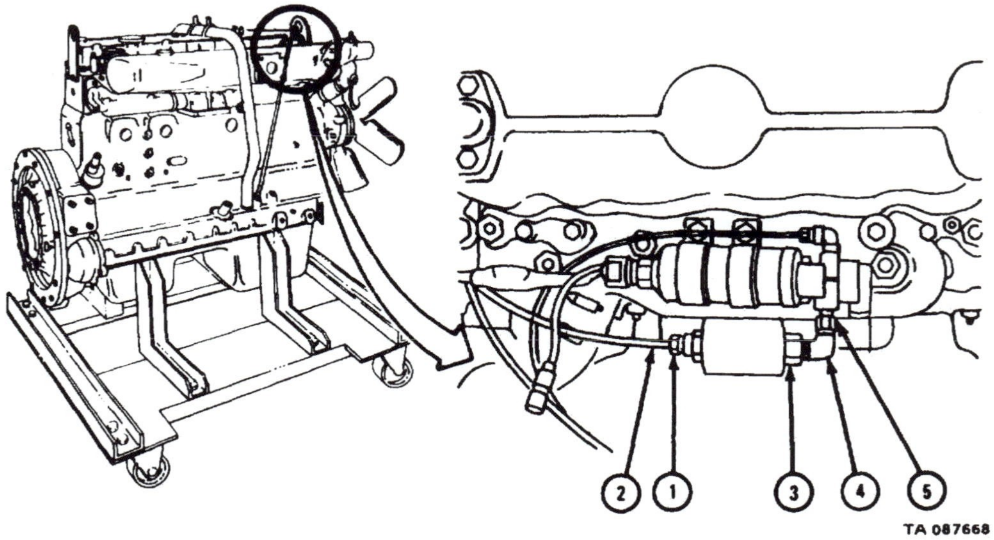

FRAME 3

NOTE

Tag all tubes and wires so that they will be put back in the same place.

1. Unscrew nut (1) and take out flame heater fuel pump-to-flame heater nozzle tube (2).
2. Unscrew and take out elbow (3).
3. Unscrew nut (4) and take out flame heater wiring harness (5).
4. Take out two screws and lockwashers (6) and take out flame heater fuel pump (7).

NOTE

Some engines with high profile water manifolds have 0.620-inch spacers under two clamps (8). These should be taken out with fuel pump (7).

END OF TASK

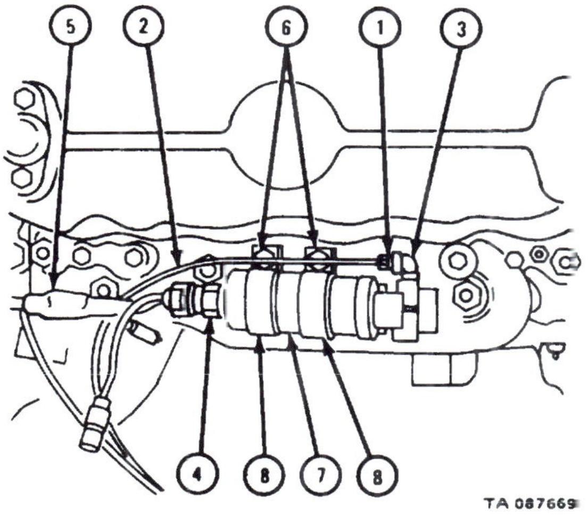

e. Side-Mounted Flame Heater Fuel Pump.

NOTE

Tag all lines so that they are put back in the same place.

FRAME 1

1. Unscrew nut (1) from fitting (2), and take out flame heater fuel pump-to-flame heater nozzle tube (3).
2. Unscrew nut (4) from elbow (5) and take out flame heater fuel return-to-solenoid valve tube (6) .

GO TO FRAME 2

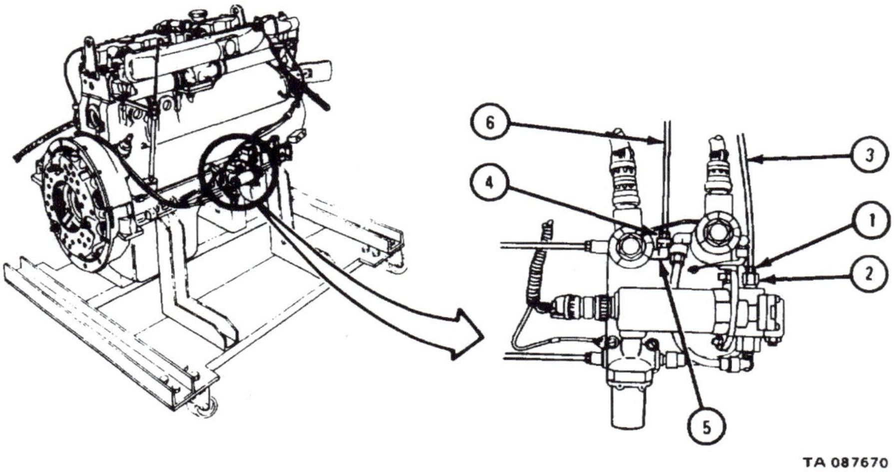

FRAME 2

1. Unscrew nut (1) from fitting (2) on fuel return solenoid valve (3) and take out fuel return solenoid valve-to- fuel injector nozzle tube (4).

2. Unscrew nut (5) from fitting (6) on fuel filter (7) and take out fuel injection pump-to-filter tube (8) .

GO TO FRAME 3

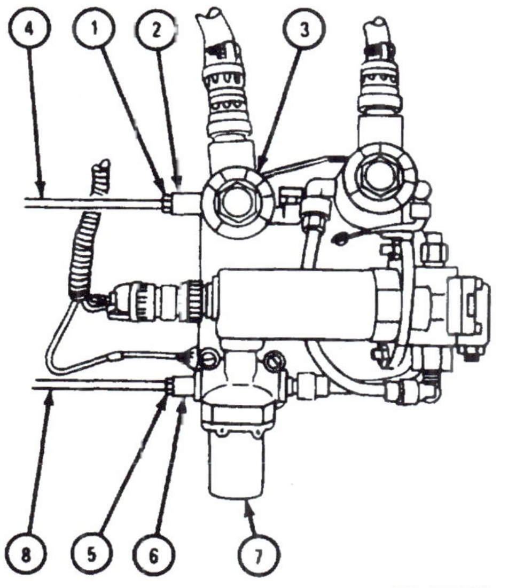

FRAME 3

1. Unscrew connector (1) from fuel supply solenoid valve (2) and pull connector clear of valve.
2. Unscrew connector (3) from fuel return solenoid valve (4) and pull connector clear of valve.
3. Unscrew connector (5) from fuel pump (6) and pull connector clear of pump.
4. Take out screw, nut, and lockwasher (7) and take off ground lead (8). Take out flame heater wiring harness (9).

GO TO FRAME 4

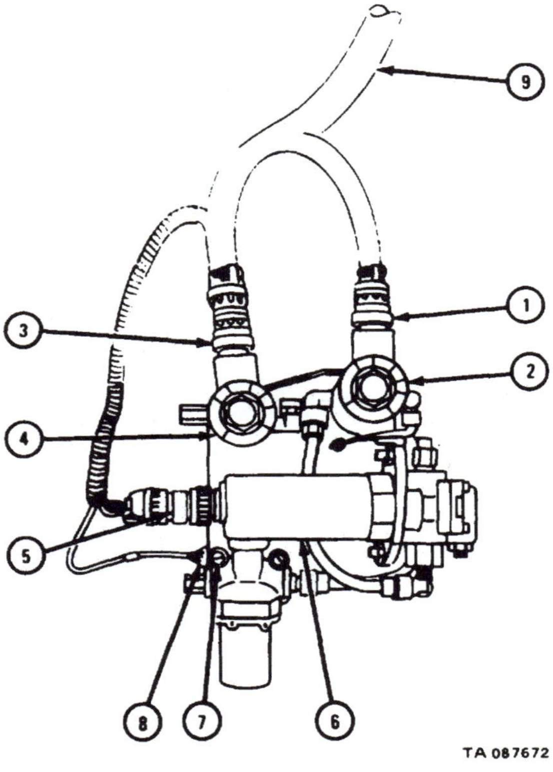

FRAME 4

1. Take out three bolts (1) and three serrated lockwashers (2) holding bracket (3) to crankcase (4). Hold up bracket so that it does not drop when taking out last bolt.
2. Carefully take out bracket (3) with fuel pump, fuel filter, and solenoid valve assembly (5).

END OF TASK

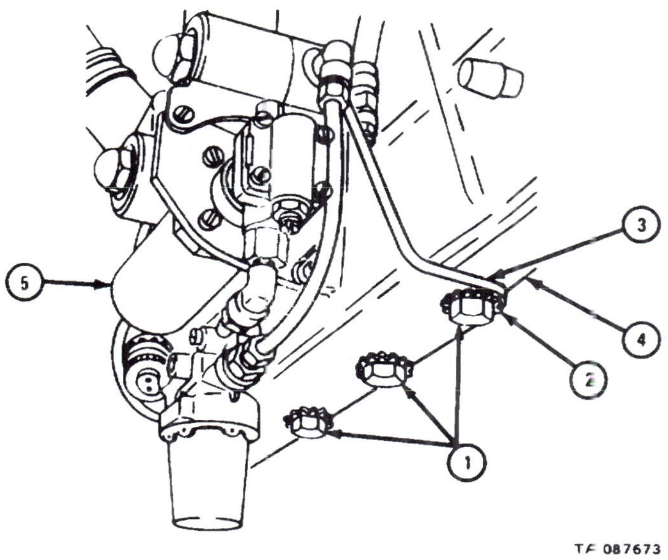

3-22. AIR PRESSURIZATION TUBE ASSEMBLIES (ENGINE LDS-465-2).

FRAME 1

NOTE

Tag air pressurization tube (2) so that it is put back in the same place.

1. Unscrew nut (1) and take off air pressurization tube (2).
2. Take off nut and flat washer (3) and take off tube clamp (4).

GO TO FRAME 2

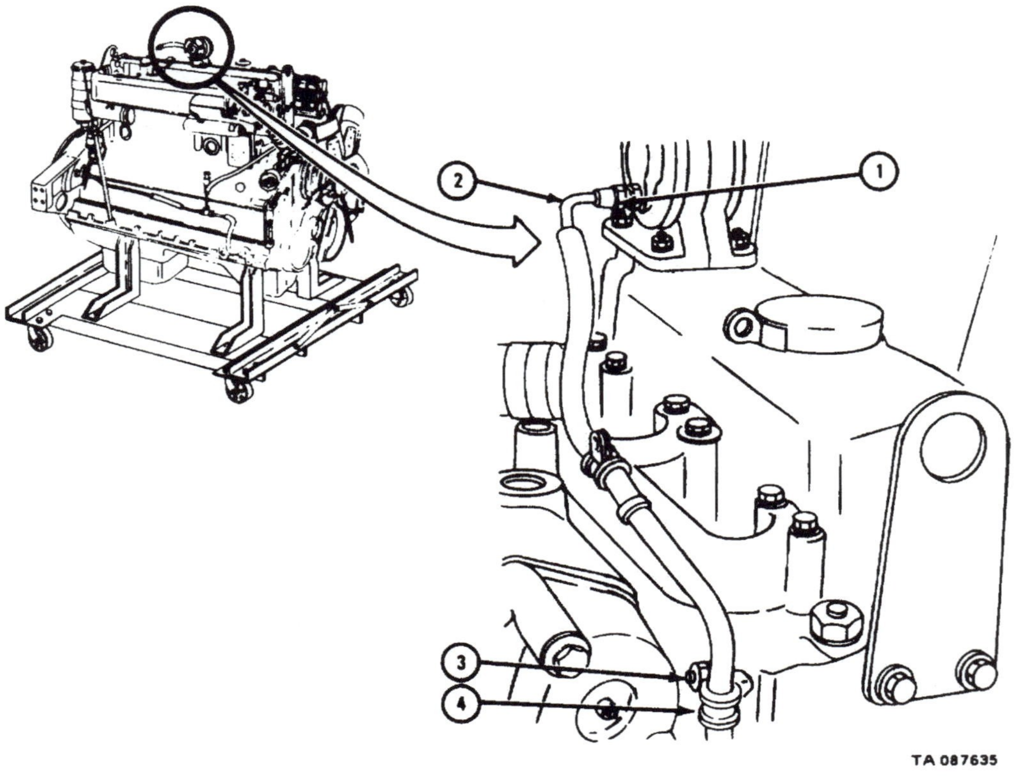

FRAME 2

NOTE

Tag all tubes so that they will be put back in the same place.

1. Unscrew nut (1) and take out air pressurization tube (2).
2. Unscrew two nuts (3) and take out flywheel housing pressurization tube (4).
3. Unscrew and take out flywheel housing tee fitting (5).

GO TO FRAME 3

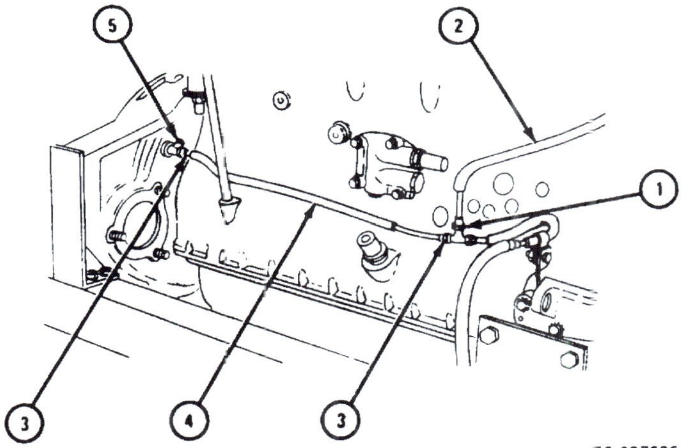

FRAME 3

NOTE

Tag all tubes so that they are put back in the same place.

1. Unscrew nut (1) and take out hose (2).
2. Unscrew nut (3), take off hose (4), and take out tee fitting (5).
3. Take out locknut (6) and screw (7) . Take out clamp (8) with hose (4) .

END OF TASK

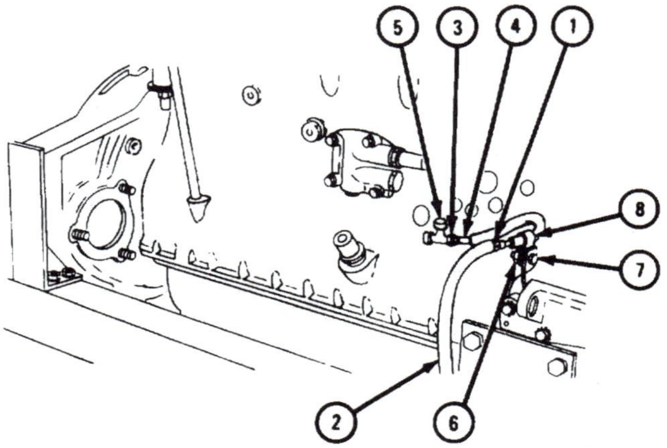

TA 087637

3-23. GENERATOR MOUNTING BRACKET.

FRAME 1

1. Take out three screws and serrated lockwashers (1).
2. Take off generator mounting bracket (2).
3. If working on engine LDS-465-2, take off air inlet tube support bracket (3).

END OF TASK

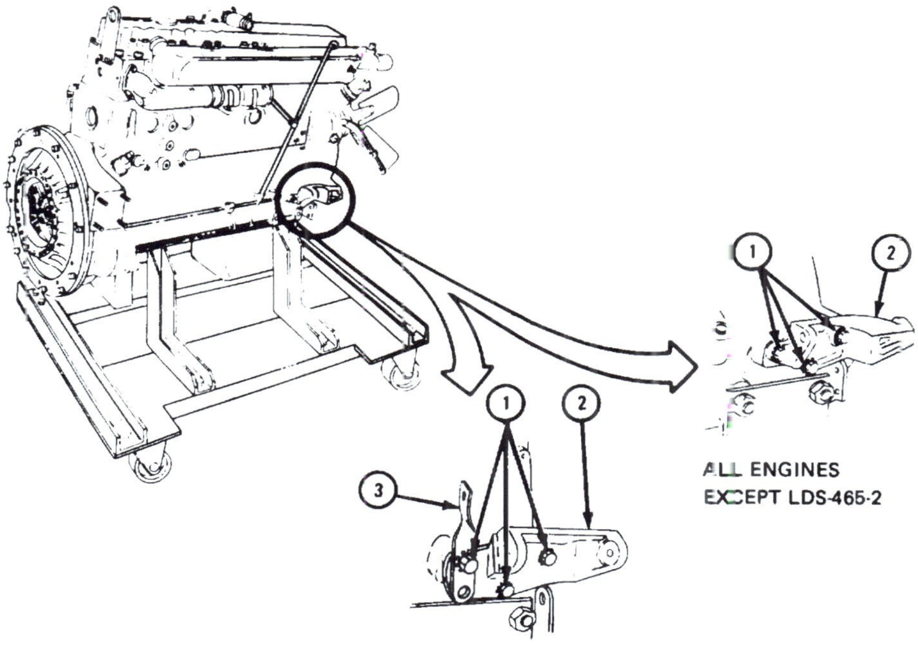

3-24. OIL PRESSURE REGULATOR VALVE HOUSING ASSEMBLY.

FRAME 1

NOTE

This task is shown for engine LD-465-1. This task is the same for other engines.

1. Take out four screws and lockwashers (1).
2. If working on engines LD-465-1 or LD-465-1C, take out crankcase breather tube support bracket (2).
3. Take off oil pressure regulator valve housing (3). Take off and throw away oil pressure regulator valve housing gasket (4).

END OF TASK

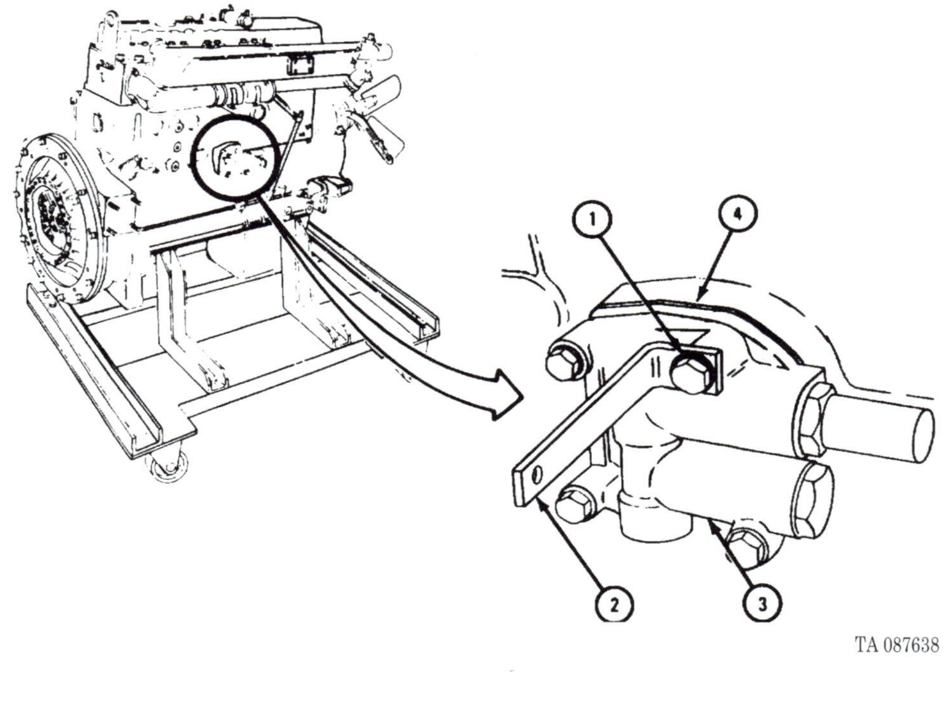

TA 087638

3-25. OIL GAGE SUPPORT.

a. Engines LD-465-1, LD-465-1C, and LDT-465-1C.

FRAME 1

1. Take out oil level gage rod (1).
2. Takeout nut, screw, and lockwasher (2).
3. Take out oil level gage rod support tube (3).

END OF TASK

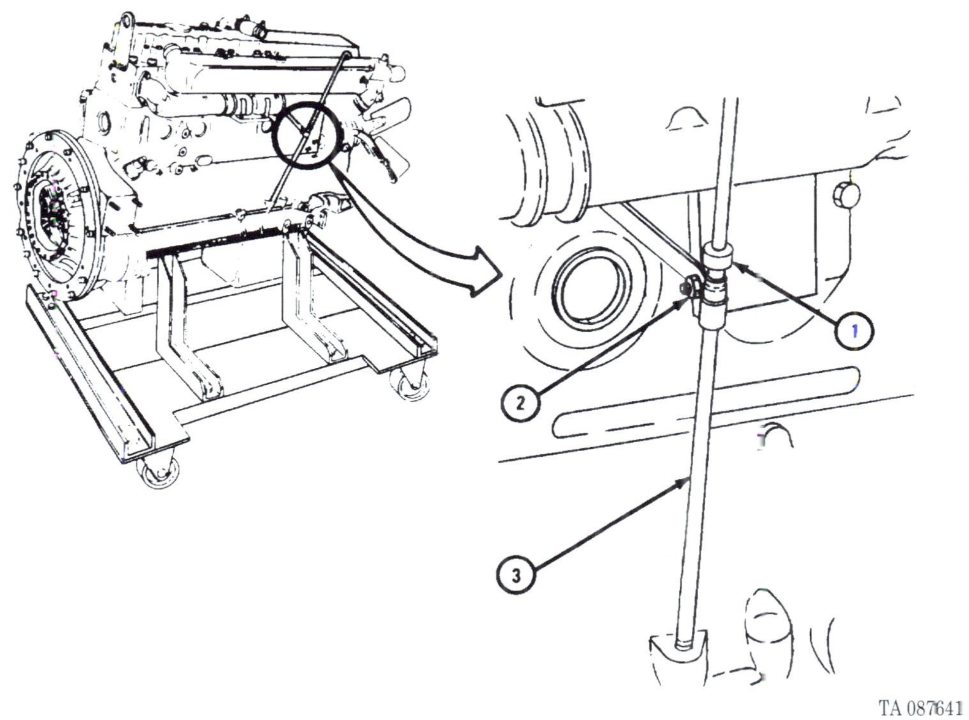

b. Engines LDS-465-1, LDS-465-1A, and LDS-465-2.

NOTE

This task is shown for engine LDS-465-1A. This task is the same for engines LDS-465-1 and LDS-465-2.

1. Unscrew and takeout oil level gage rod (1).
2. Take out nut, screw, and lockwasher (2).
3. Using pipe wrench, take out oil level gage rod support tube (3).

END OF TASK

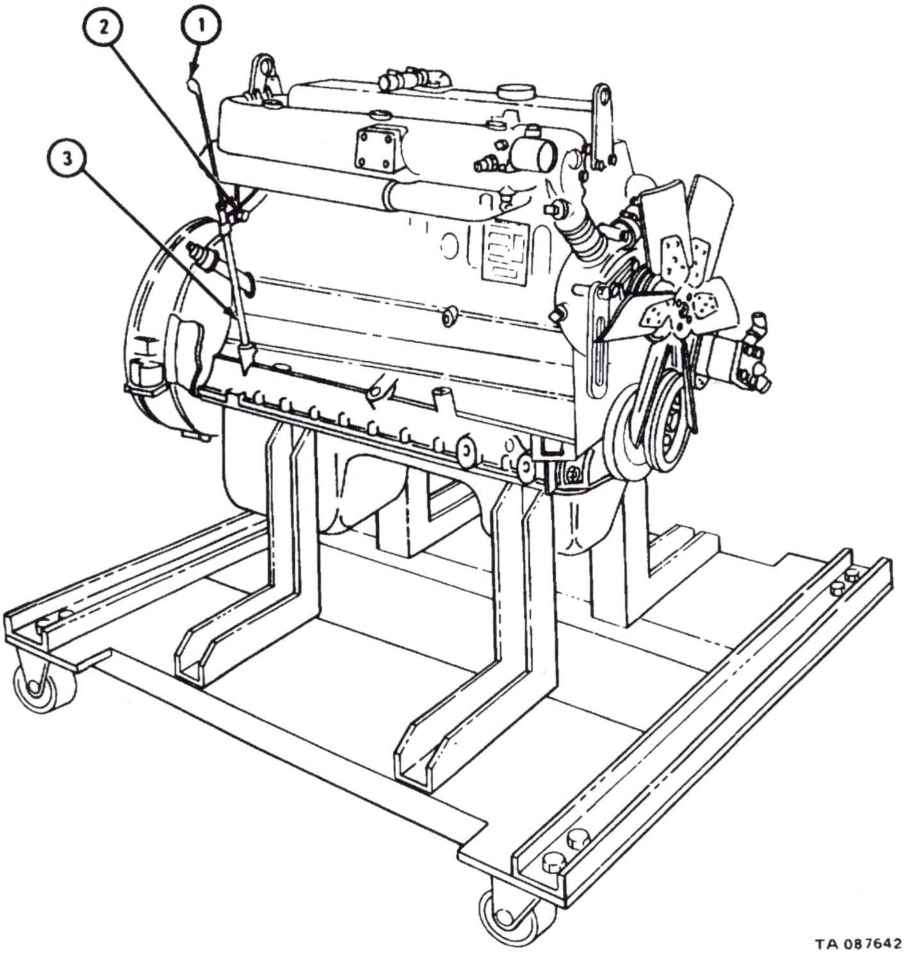

3-26. OIL PRESSURE TRANSMITTER AND COOLANT DRAIN COCK.

FRAME 1

NOTE

This task is shown for engine LD-465-1C. This task is the same for all other engines.

Oil pressure transmitter (1) may have already been taken off engine LDS-465-1A when engine was taken out of truck.

1. Hold pipe adapter (2). Take out oil pressure transmitter (1).
2. Take out pipe adapter (2).

NOTE

Engine LDS-465-2 has 90° elbow (3) instead of 45° elbow shown.

3. Take out elbow (3).
4. Take out coolant drain cock (4).
5. Take out elbow (5).

END OF TASK

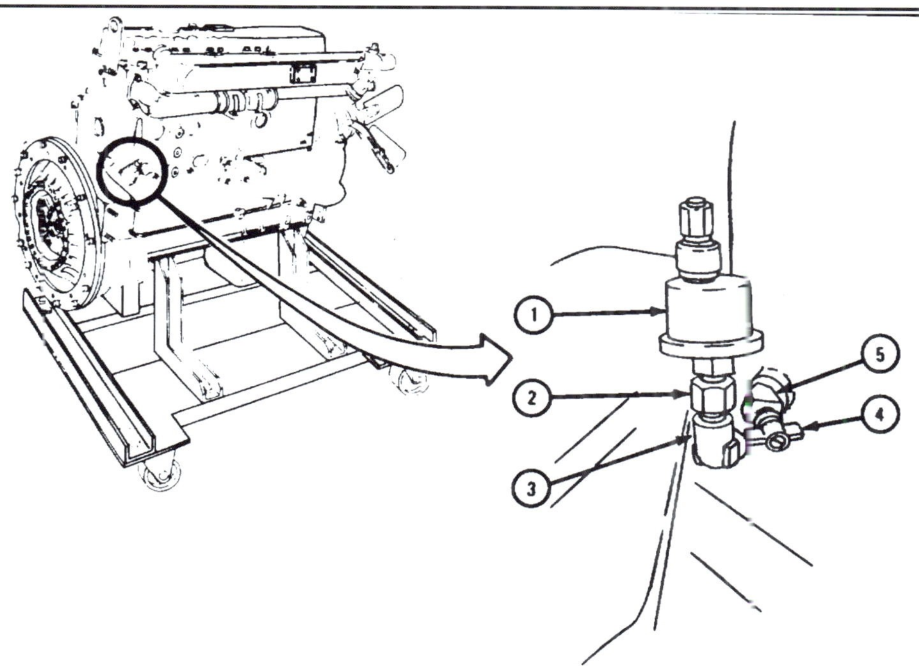

TA 087640

3-27. ENGINE COOLING FAN, THERMOSTAT, AND WATER PUMP ASSEMBLY.

a. Engines LD-465-1, LD-465-1C, LDT-465-1C, LDS-465-1, and LDS-465-1A.

(1) Engine coolant fan.

FRAME 1

1. Takeout four screws and lockwashers (1).
2. Takeoff fan (2).

IF WORKING ON ENGINE LDS-465-1, GO TO FRAME 2.
IF WORKING ON ENGINES LD-465-1, LD-465-1C, LDT-465-1C, OR LDS-465-1A, END OF TASK

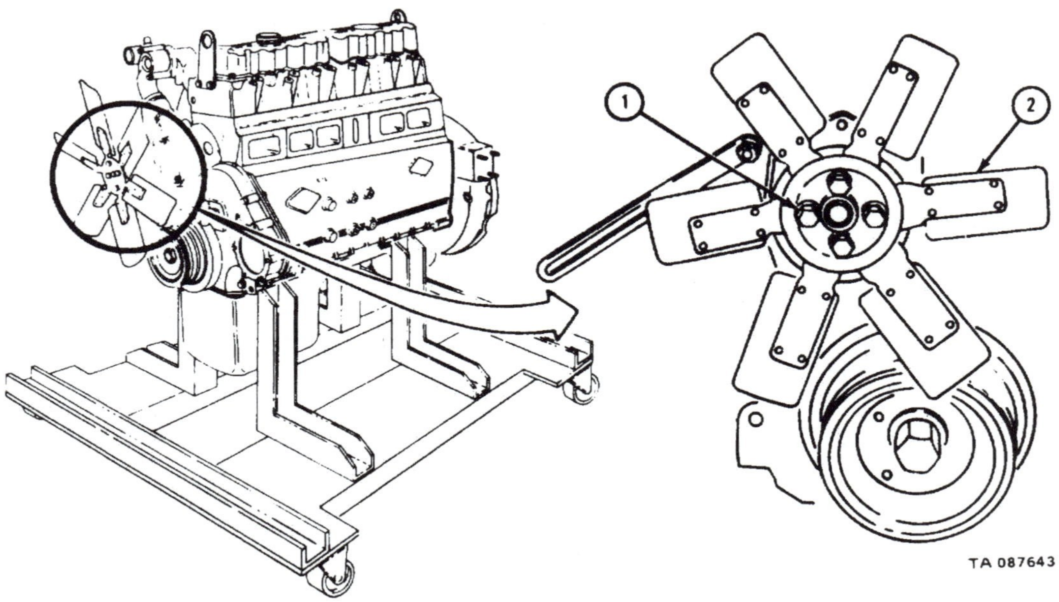

FRAME 2

1. Take out engine fan spacer plate (1).

END OF TASK

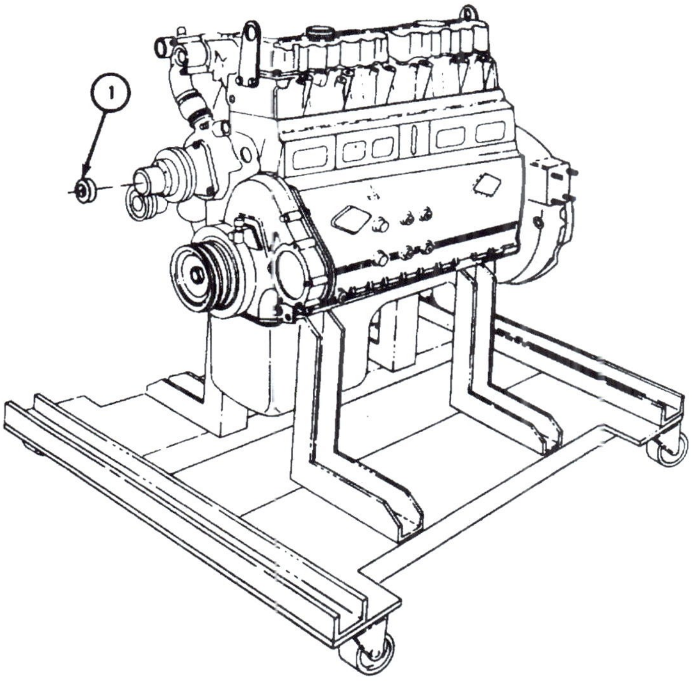

(2) Thermostat housing.

FRAME 1

NOTE

On some engines, generator adjusting strap (2) is curved instead of straight as shown. This task is the same for all generator adjusting straps.

1. Take out nut and lockwasher (1) and take off generator adjusting strap (2).
2. Loosen hose clamp (3).

GO TO FRAME 2

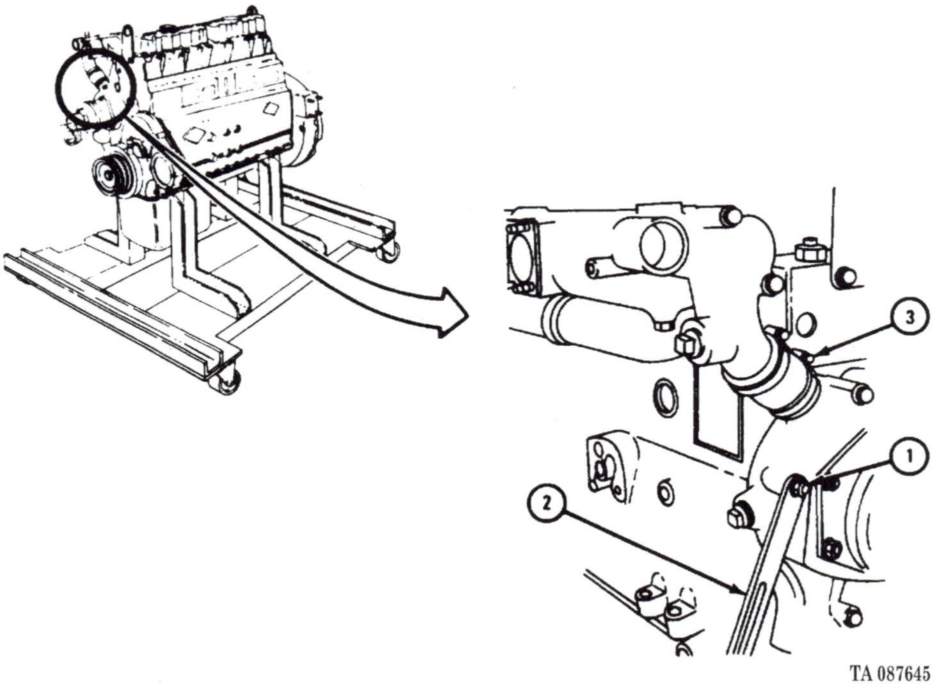

FRAME 2

1. Take out two screws and flat washers (1).
2. Takeout thermostat housing (2) with hose (3) .
3. Take off thermostat housing gasket (4) and threw it away.

END OF TASK

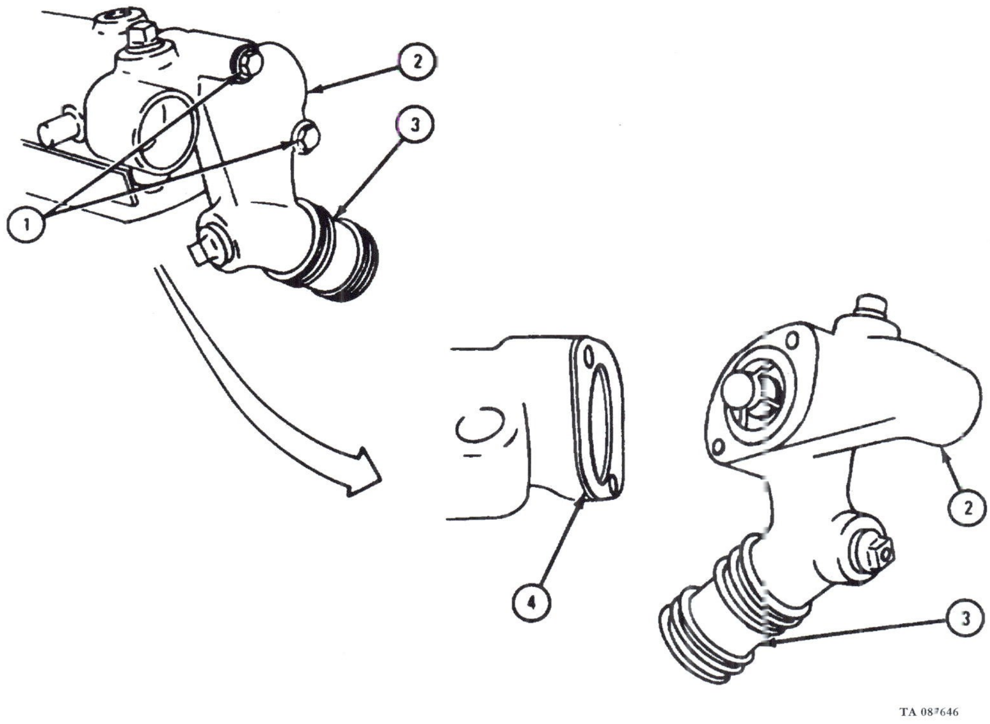

(3) Water pump assembly .

FRAME 1

NOTE

Early LDS-465-1 engine water pump assemblies (2) have mounting flange (3) instead of tube (4). This task i s the same for both types of pump.

1. Take out . three screws and lockwashers (1) and take off water pump assembly (2).

END OF TASK

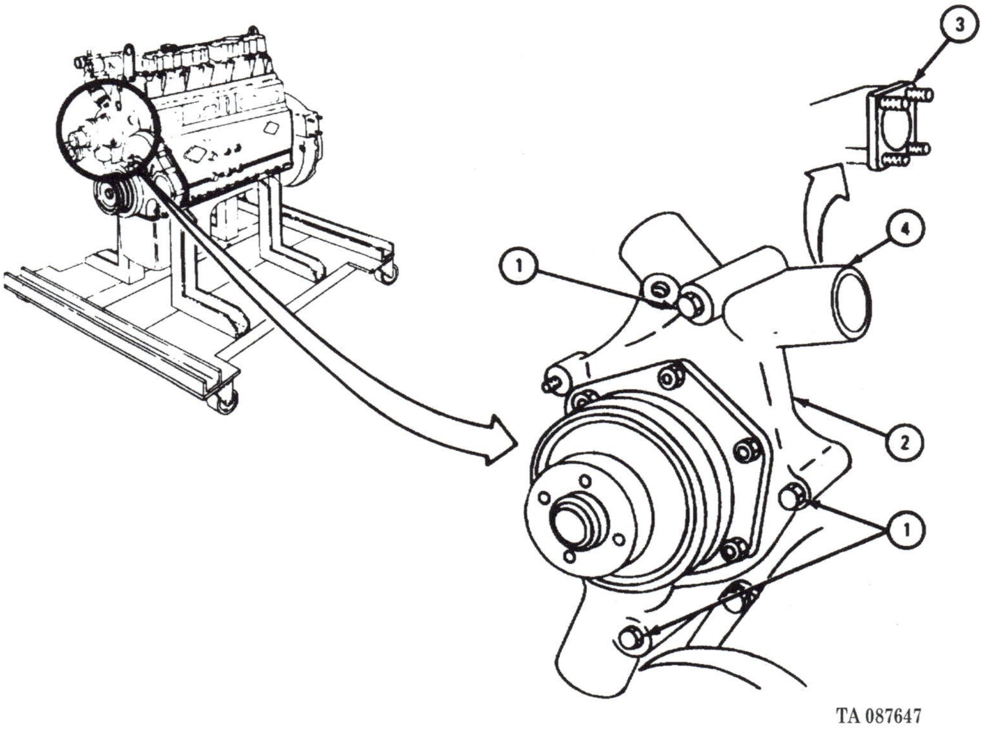

b. Engine LDS-465-2 .

(1) Engine cooling fan .

FRAME 1

1. Cut and take out three lockwires (1).
2. Take out six screws and lockwashers (2) and take off fan (3).

END OF TASK

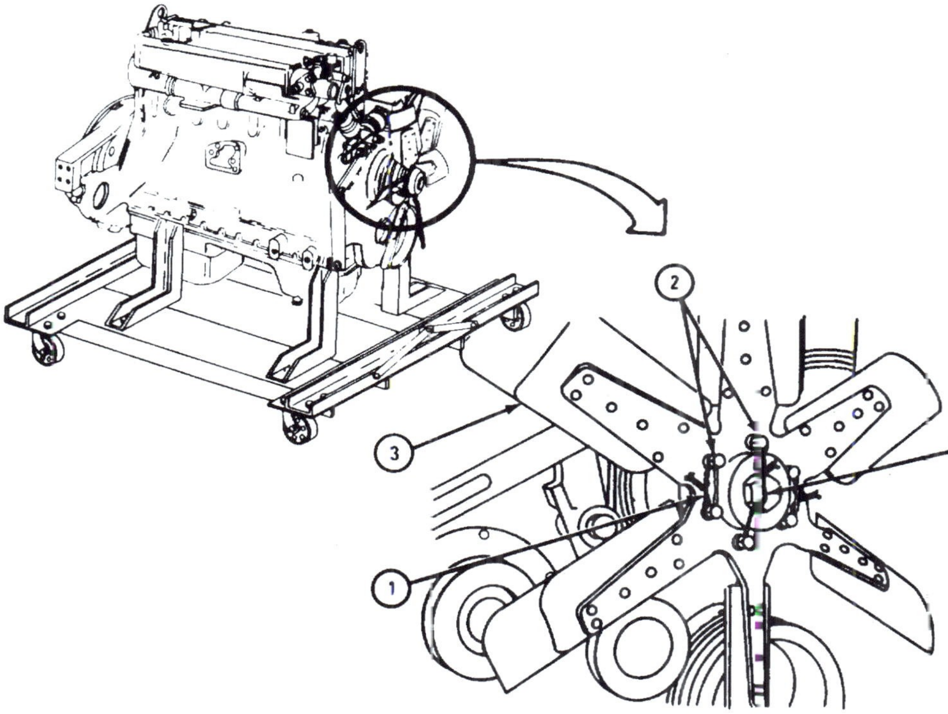

TA 087348

(2) Thermostat housing .

FRAME 1

1. Takeout screw, lockwasher, and serrated washer (1) .
2. Takeoff nut and lockwasher (2) and takeout idler pulley adjusting strap (3) and generator adjusting strap (4) .
3. Take out screw, lockwasher, and plain washer (5) and take out fan idler pulley adjusting arm (6).
4. Take off two fan drive belts (7) .

GO TO FRAME 2

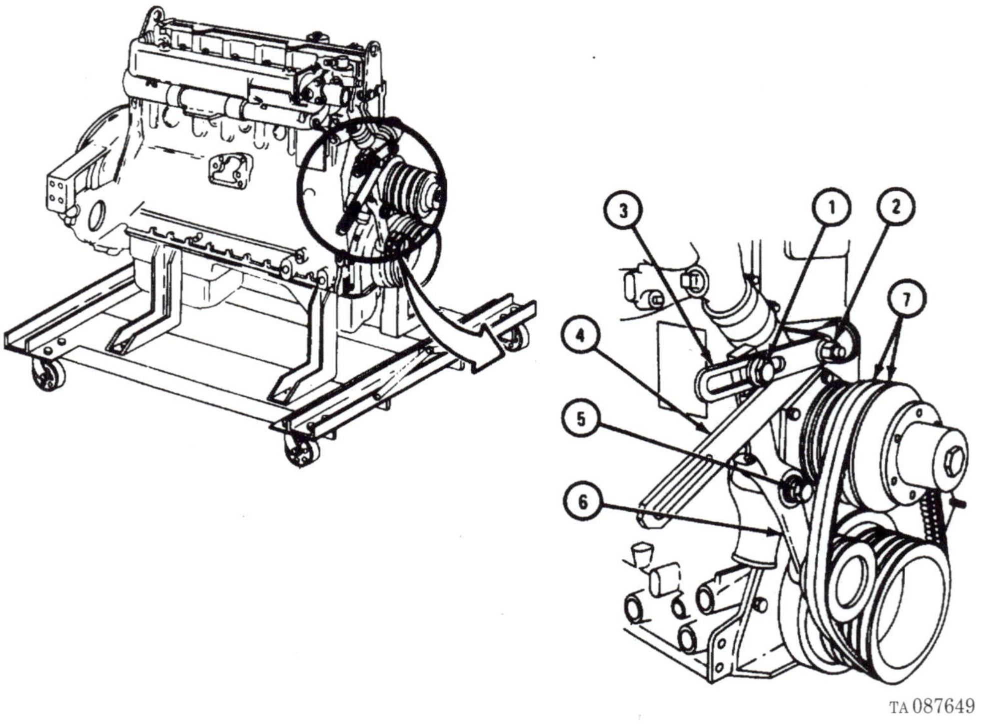

FRAME 2

1. Loosen hose clamp (1).
2. Take out two screws and flat washers (2).
3. Take out thermostat housing (3) with hose (4).
4. Take off thermostat housing gasket (5) and throw it away.

END OF TASK

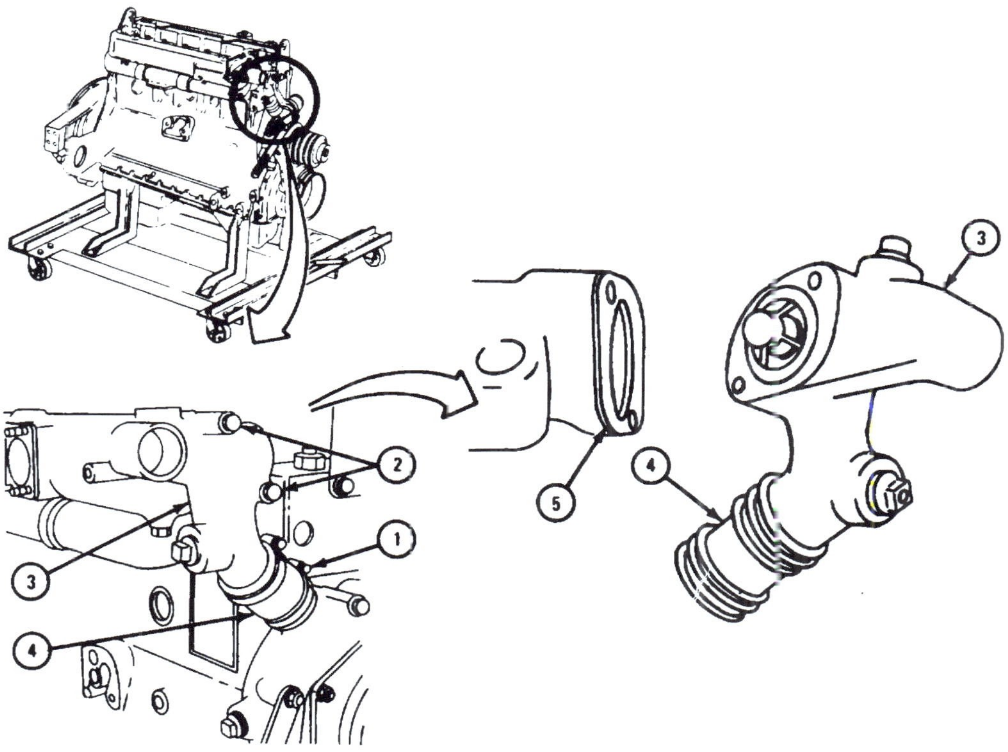

TA 087650

(3) Water pump assembly .

FRAME 1

1. Take out six nuts and lockwashers (1) and take off water pump (2).
2. Take out three screws and lockwashers (3) and take off water pump housing (4) .

END OF TASK

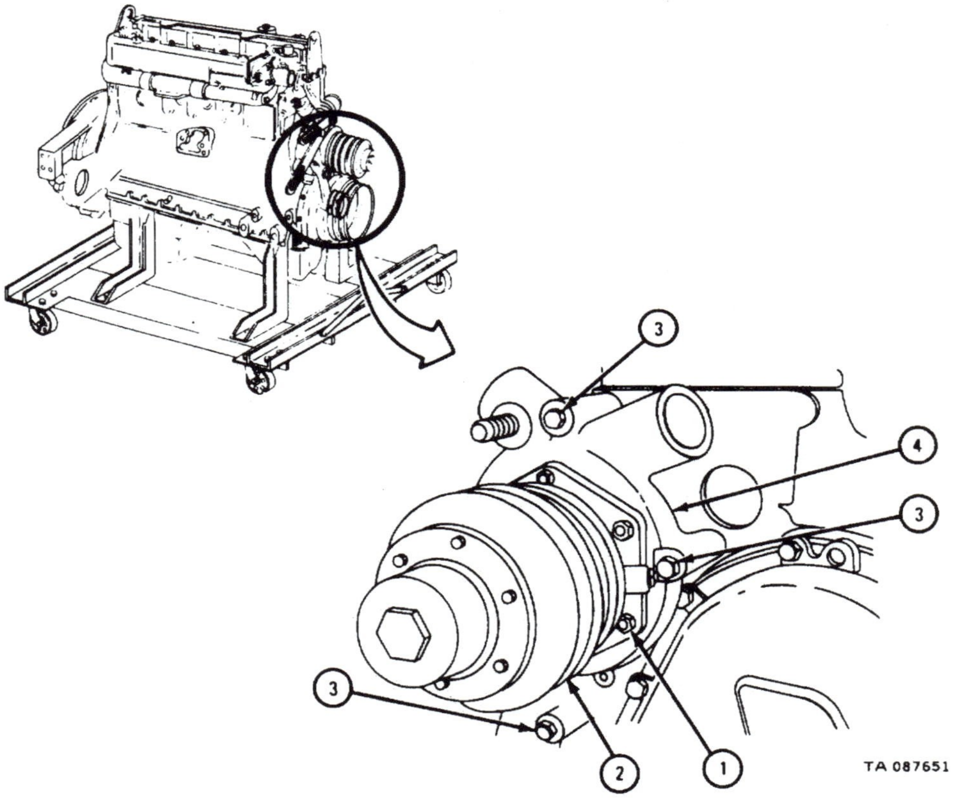

3-28. CYLINDER HEAD COVER.

a. All Engines Except LDS-465-2.

FRAME 1

1. Take off locknut and flat washer (1).
2. Take out locknut (2) and takeoff tachometer cable bracket (3).

NOTE

Engine LDS-465-1 does not have chain (4).

GO TO FRAME 2

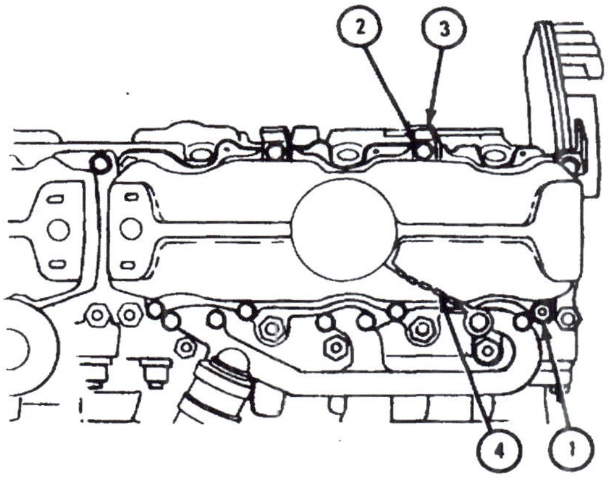

TA 037655

FRAME 2

1. Take off oil filler cap (1).
2. Take out four lockbolts and washers (2) and cover (3) . take off front cylinder head .
3. Take off cylinder head cover gasket (4) and throw it away.

GO TO FRAME 3

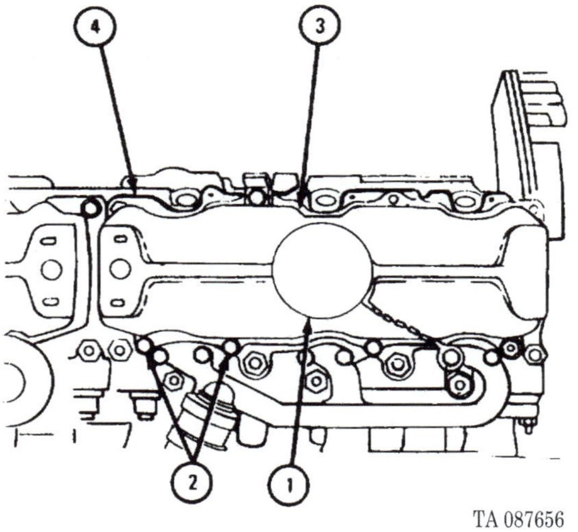

TA 087656

FRAME 3

1. Take off locknut and washer (1).
2. Take out lockbolt and washer (2) and take off tachometer cable mounting bracket (3).
3. Take out six lockbolts and washers (4) and take off rear cylinder head cover (5).
4. Take off rear cylinder head cover gasket (6) and throw it away.

END OF TASK

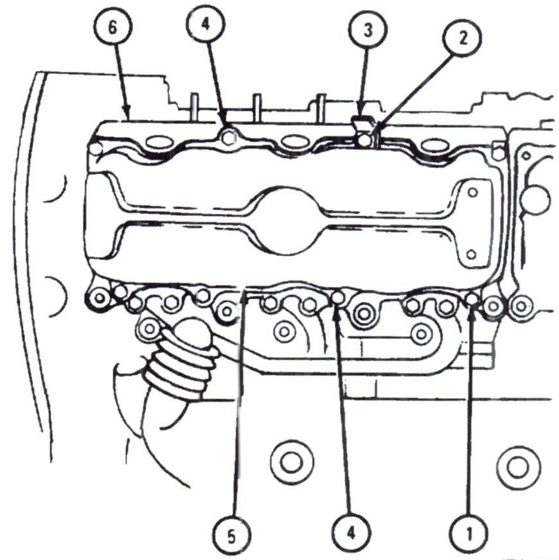

TA 0876-8

b. Engine LDS-465-2.

FRAME 1

1. Take off oil filler cap (1).
2. Take off locknut and flat washer (2).
3. Take out seven lockbolts and plain washers (3) and take off front cylinder head cover (4).
4. Take off front cylinder head cover gasket (5) and throw it away.

GO TO FRAME 2

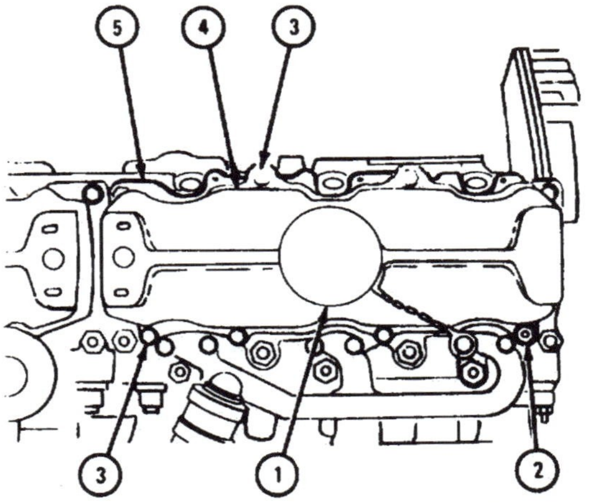

TA 087734

FRAME 2

1. Take off locknut and washer (1).
2. Take out seven lockbolts and washers (2) and take off rear cylinder head cover (3).
3. Take off rear cylinder head cover gasket (4) and throw it away.

END OF TASK

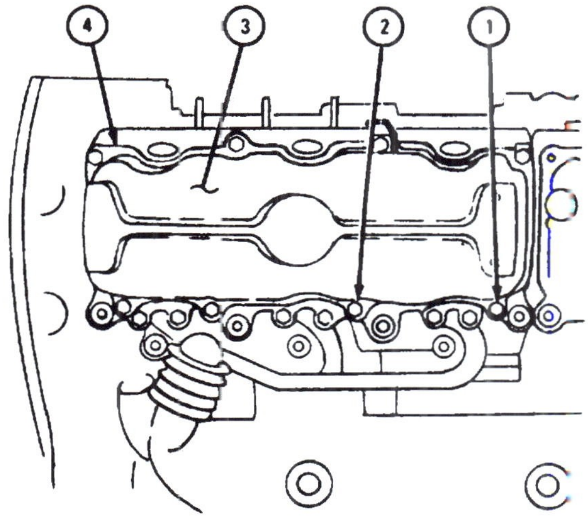

TA 087745

3-29. CYLINDER HEAD WATER OUTLET MANIFOLD.

FRAME 1

1. Loosen two hose clamps (1).

NOTE

Three of 12 screws (2) may already have been taken out.

2. Take out 12 screws (2).

NOTE

Do step 3 for all engines except LDS-465-1.

3. Take off chain retainer and oil filler cap assembly (3).
4. Take off front water outlet manifold (4) and rear water outlet manifold (5).
5. Take off and throw away six water outlet manifold gaskets (6).

END OF TASK

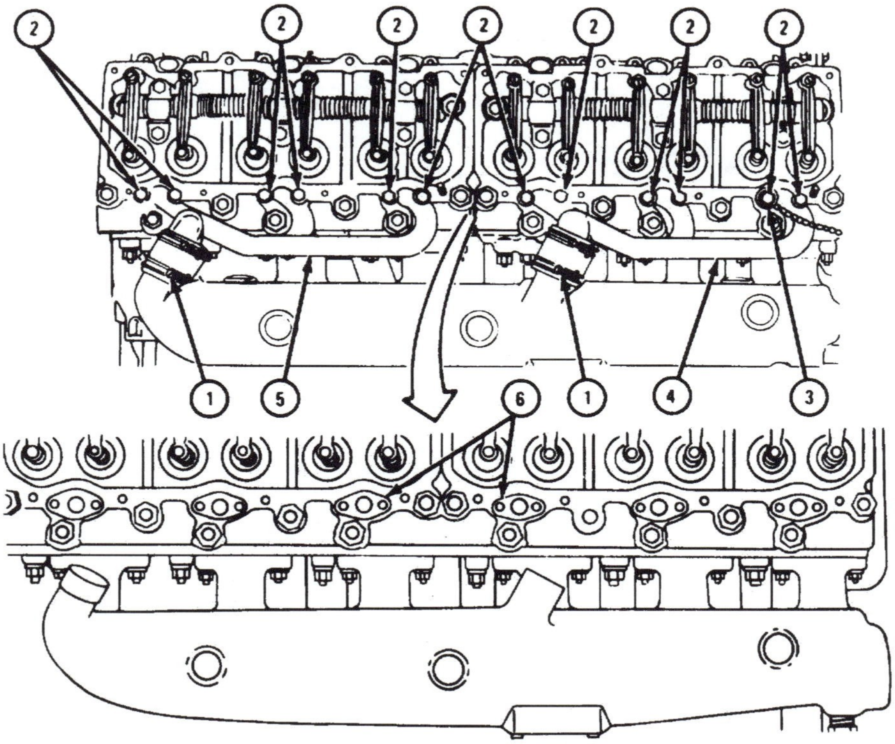

TA 087674

3-30. INTAKE AND EXHAUST MANIFOLDS.

a. Preliminary Procedure . Remove cylinder head water outlet manifolds. Refer to para 3-29.

b. Removal.

NOTE

Intake manifold and exhaust manifold must be taken off together as an assembly.

FRAME 1

1. Take out 12 nuts and washers (1) holding top flanges of intake manifold (2) to cylinder heads (3) .
2. Take out six locknuts and washers (4) holding top flanges of exhaust manifold (5) to cylinder heads (3) .

IF WORKING ON ENGINES LDS-465-1 , LDS-465-1A , OR LDS-465-2, GO TO FRAME FRAME 2.
IF WORKING ON ENGINES LD-465-1, LD-465-1C , OR LDT-465-1C , GO TO FRAME 3

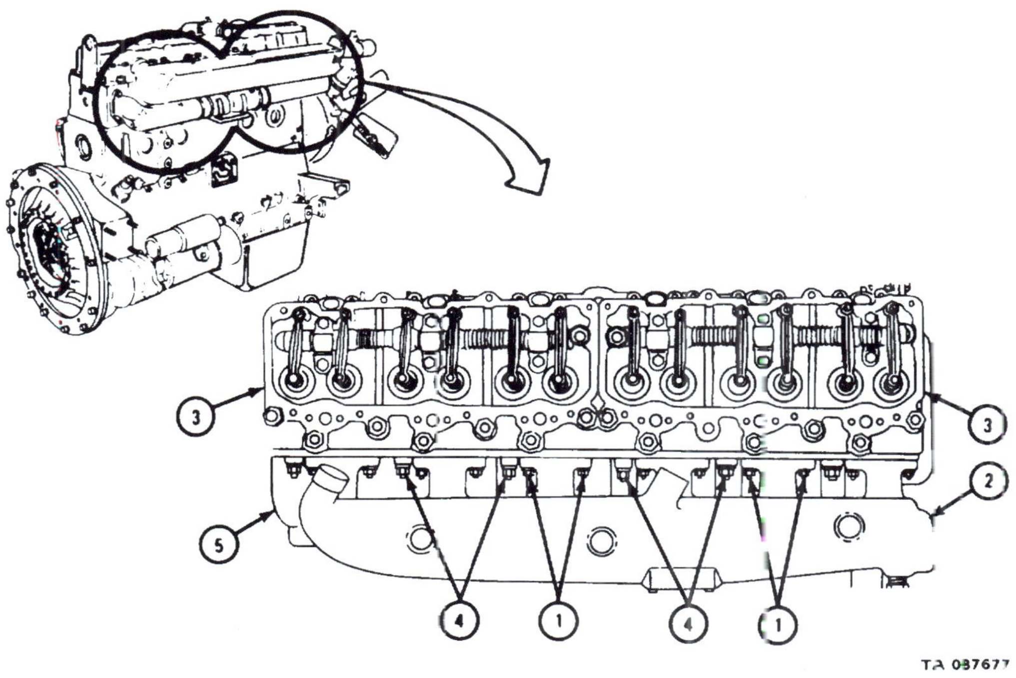

FRAME 2

1. Take off six nuts and washers (1) holding bottom flanges of intake manifold (2) to cylinder heads (3) .
2. Take off two locknuts (4) and take off oil level gage support bracket (5).

WARNING

Intake manifold (2) and exhaust manifold (7) assembly is heavy. Hold it up when taking off locknuts and washers (6) or it may fall and cause injury to personnel.

3. Take off four locknuts and washers (6) holding exhaust manifold (7) to cylinder heads (3).
4. Take off intake manifold (2) and exhaust manifold (7) together as an assembly.
5. Take intake manifold (2) apart from exhaust manifold (7).

GO TO FRAME 4

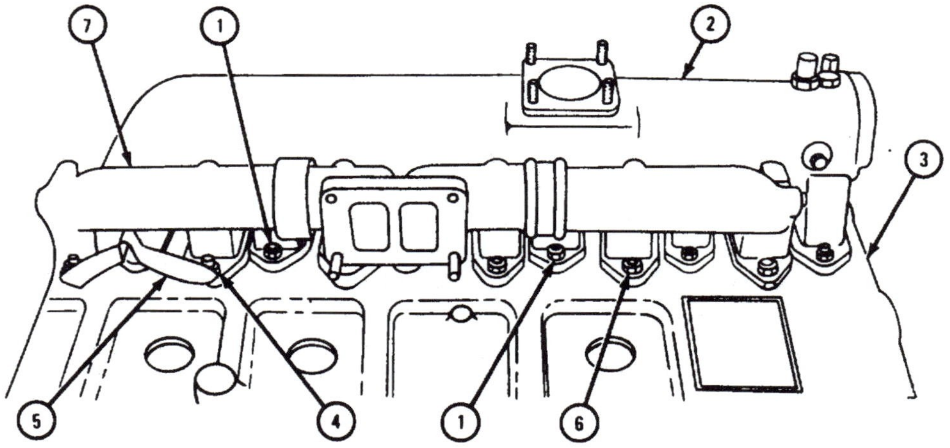

TA 087678

FRAME 3

1. Take off four nuts and washers (1) holding bottom flanges of intake manifold (2) to cylinder heads (3) .
2. Take off two nuts (4) and take off oil level gage support bracket (5).

WARNING

Intake manifold (2) and exhaust manifold (7) assembly is heavy. Hold it up firmly when taking off locknuts and washers (6) or it may fall and cause injury to personnel.

3. Take off six locknuts and washers (6) holding exhaust manifold (7) to cylinder head (3) .
4. Take off intake manifold (2) and exhaust manifold (7) together as an assembly.
5. Take intake manifold (2) apart from exhaust manifold (7).

GO TO FRAME 4

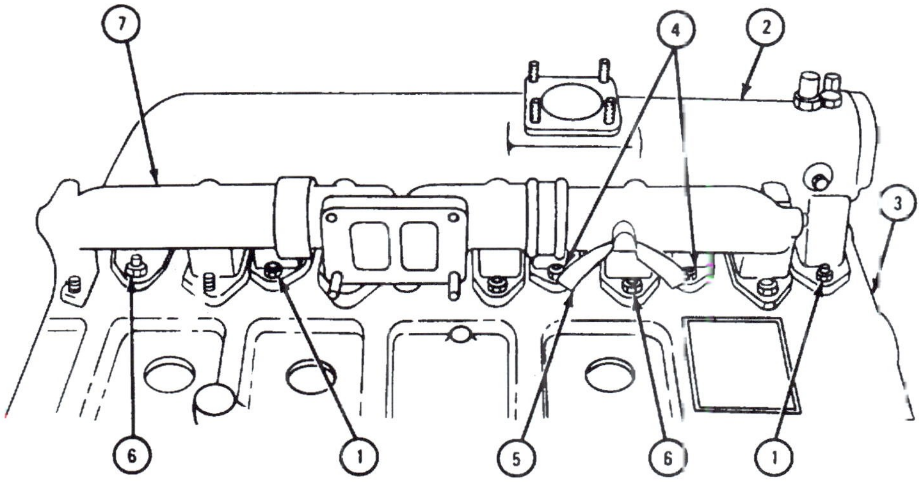

FRAME 4

1. Takeoff two manifold gaskets (1) and them them away .

END OF TASK

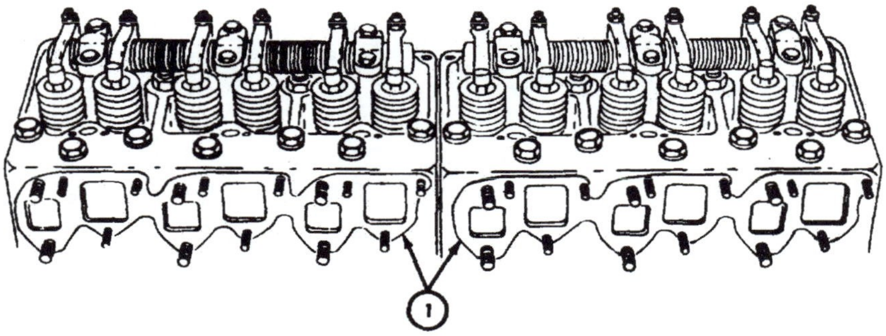

TA 087680

Section III. DISASSEMBLY OF ENGINE ASSEMBLY

NOTE

This procedure is the same for all engine models except where noted.

TOOLS: Engine transport stand, NSN 4910-00-338-6673
Engine maintenance stand, NSN 4910-00-529-8387
Puller, NSN 5120-00-134-7463
Puller kit, pn 8708724
Plug, pn 10899178
Wood block, 2x4x18 inches (2)
Brass rod

SUPPLIES: Lacquer, nitrocellulose, type I, TT-L-50
Tool tags
Wire
Steel bar stock
Bolt (2)
Tags

PERSONNEL: Two

EQUIPMENT CONDITION: Engine mounted on transport stand.

3-31. MOUNTING ENGINE ON MAINTENANCE STAND.

a. Preliminary Procedures.

(1) Remove engine from shipping container and place on transport stand. Refer to para 2-3 through para 2-5.

(2) Remove generator and mounting bracket. Refer to para 3-4.

(3) Remove crankcase breather adapter. Refer to para 3-19.

(4) If engine is equipped with a turbocharger, remove turbocharger. Refer to para 3-5.

(5) Remove oil pressure regulator valve. Refer to para 3-24.

(6) Remove air compressor and mounting bracket. Refer to para 3-12 and 3-17.

(7) Remove fuel filter and mounting bracket. Refer to para 3-14.

(8) If engine is equipped with hydraulic pump, remove hydraulic pump. Refer to para 3-13.

(9) Remove fuel injector pump and fuel injector lines. Refer to para 3-10 and 3-8.

(10) Remove oil filter and mounting bracket. Refer to para 3-11 and 3-15.

(11) Remove oil cooler and oil cooler water hose. Refer to para 3-15.

(12) Remove tappet chamber covers. Refer to para 3-16.

(13) Remove intake and exhaust manifolds. Refer to para 3-30.

b. Mounting.

FRAME 1

NOTE

Engine should be mounted on stand with the center of balance of engine in the center of mounting area.

Soldiers A and B

1. Using chain hoist, lift engine (1) off transport stand.
2. Move chain hoist so engine (1) hangs between engine stand mounting plates (2) .

GO TO FRAME 2

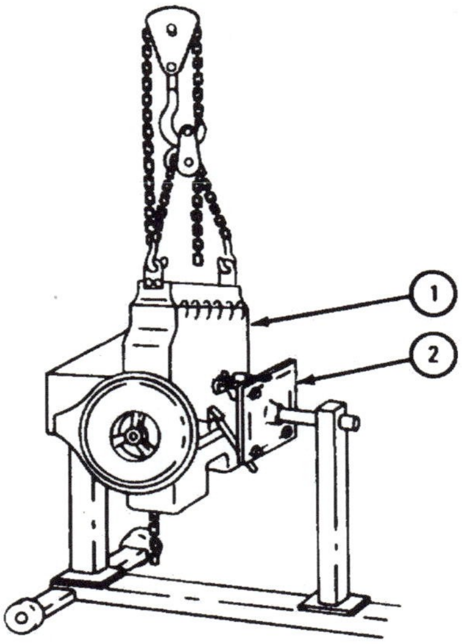

TA 118082

FRAME 2

WARNING

Do not use less than six universal mounting arms. Using less than six arms may cause the stand to fall, causing damage to equipment and injury to personnel.

NOTE

Capscrews used to mount engine accessories will be used to mount engine to maintenance stand. Universal mounting arms will be placed as far apart as practical and at various angles .

Do not tighten capscrews until all capscrews are put in place.

1. Working on crank handle side of engine stand, put two universal mounting arms (1) in place on mounting plate (2). Put in two capscrews (3).
2. Put two L-shaped brackets (4) in place on engine block (5).
3. Put in two capscrews (6) .

GO TO FRAME 3

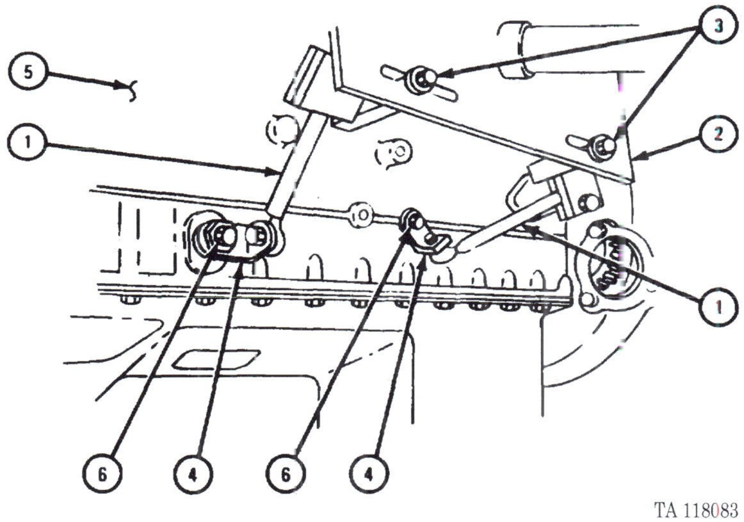

TA 118083

FRAME 3

1. Working on crank handle side of maintenance stand, put three universal mounting arms (1) on mounting plate (2). Put in three capscrews (3) .
2. Put three L-shaped brackets (4) in place on engine block (5). Put in three capscrews (6) .

GO TO FRAME 4

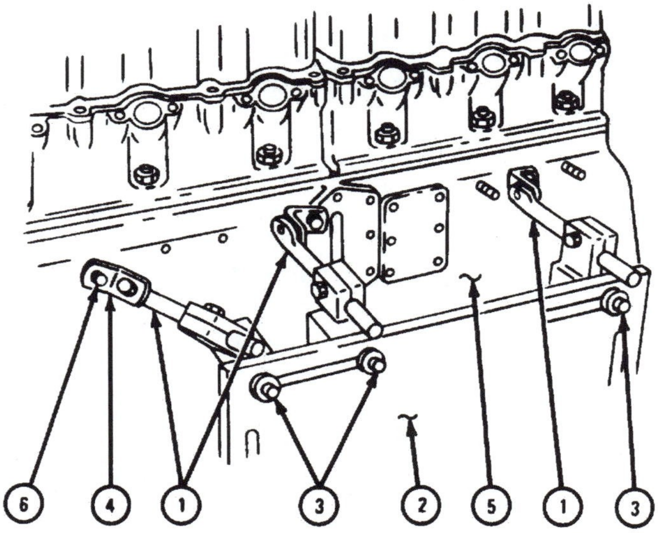

TA 118084

FRAME 4

1. Working on side of stand without crank handle, put three universal mounting arms (1) in place on mounting plate (2). Put in three capscrews (3) .
2. Put three L-shaped brackets (4) in place on engine block (5). Put in three capscrews (6) .
3. Tighten all capscrews .
4. Lower chain hoist and take off lifting sling.

END OF TASK

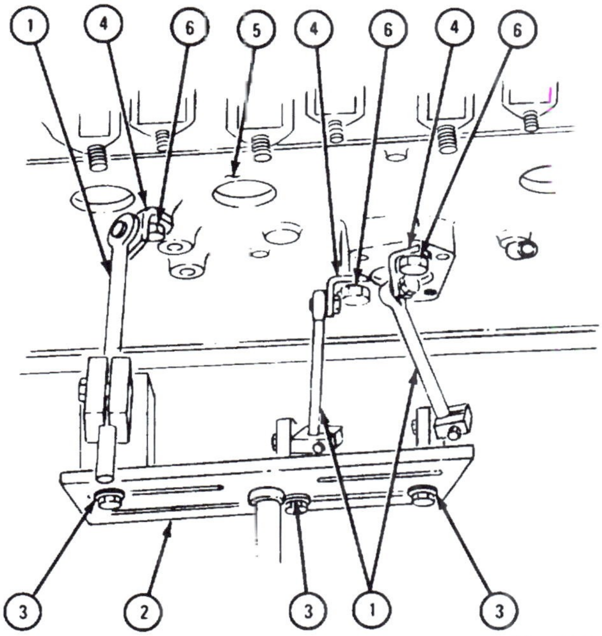

TA 118085

3-32. ENGINE FRONT AND REAR LIFTING BRACKETS .

FRAME 1

1. Take out two screws and lockwashers (1) and take off front lifting bracket (2) .
2. On all engines except engine LDS-465-2, do step 1 again for rear lifting bracket (3) .
3. On engine LDS-465-2, take out two screws and lockwashers (4) and take out rear lifting bracket and fuel filter assembly (5).

END OF TASK

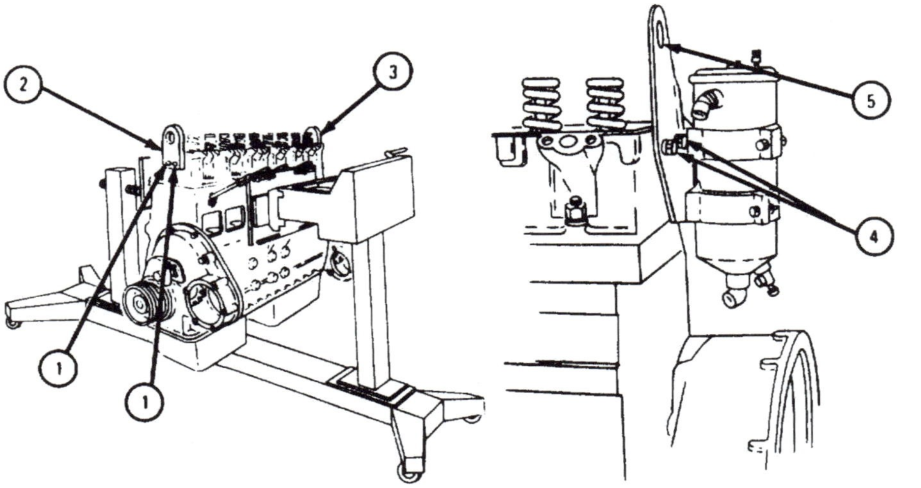

TA 087738

3-33. ROCKER ARMS, ROCKER ARM PUSH RODS, AND VALVE TAPPETS.

FRAME 1

1. Loosen 12 adjusting screw locknuts (1) and turn 12 adjusting screws (2) to the left to loosen tension on valve springs (3).
2. Take out 12 screws and lockwashers (4).
3. Take out front and rear rocker arm assemblies (5) . Tag each rocker arm assembly so it can be put back in the same place.

NOTE

> When pushrods (6) are taken out, valve tappets (7) may come out with them. If this happens, tag tappet so it can be put back in the same bore it came out of. Number valve tappets, tappet bores, and push rods (6) by numbers one to 12 starting from front to rear of engine.

4. Twist, lift out, and tag 12 push rods (6).

GO TO FRAME 2

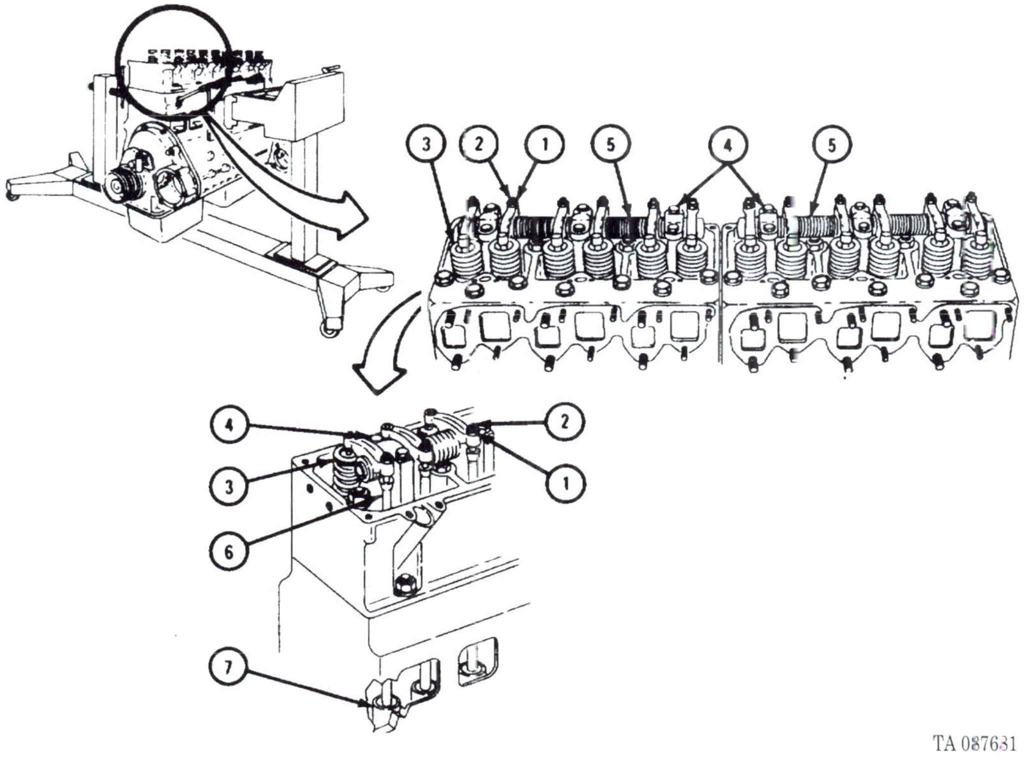

TA 087631

FRAME 2

NOTE

Tag each tappet (1) when it is taken out so it can be put back in the same place.

1. **Take out** 12 tappets (1).

END OF TASK

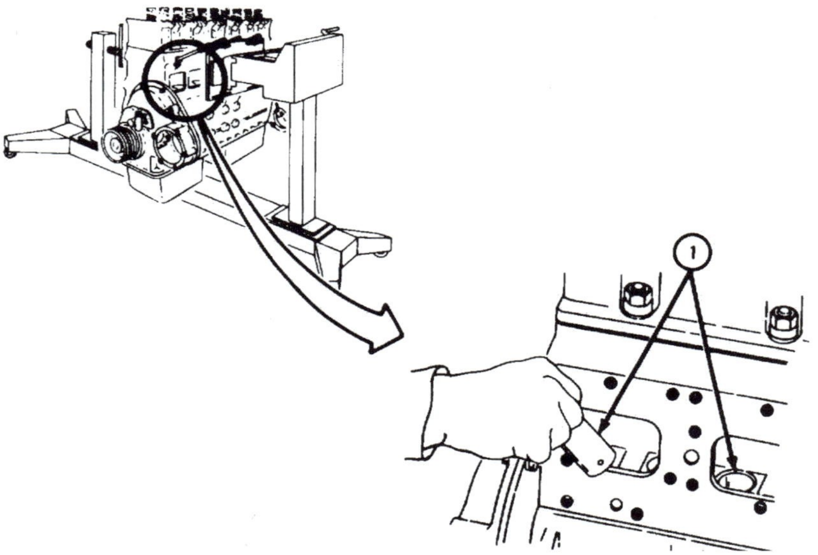

3-34. CYLINDER HEAD ASSEMBLIES.

FRAME 1

1. Take out 11 nuts and washers (1) from front cylinder head assembly (2).

NOTE

Early model engines have a sleeve spacer instead of washer (4) under each of three nuts (3) .

2. Take off three nuts (3) and three washers (4) from front cylinder head assembly (2) .
3. Do steps 1 and 2 again for rear cylinder head assembly (5).

GO TO FRAME 2

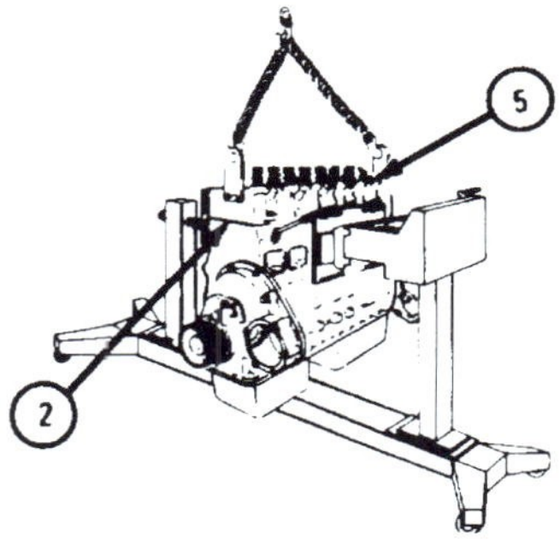

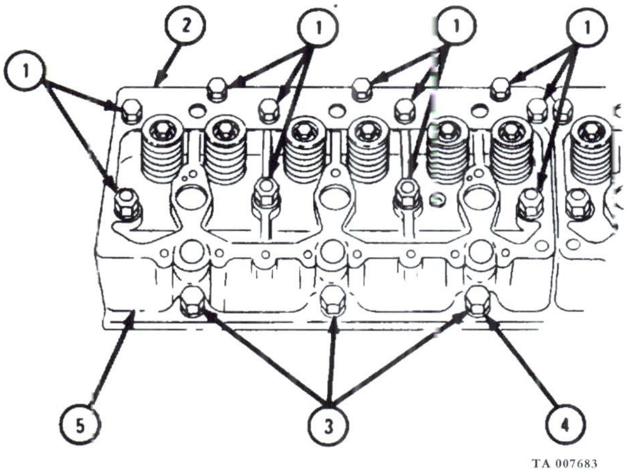

FRAME 2

1. Join lifting sling (1) to front cylinder head assembly (2) using screw and flat washer (3) and nut and flat washer (4) as shown.

CAUTION

Make sure chain hoist (5) is in position directly above cylinder head assembly (2) to be taken out. Cylinde r head assembly must be lifted straight up or mounting studs at top of engine (6) may be bent.

Soldier A 2. Hook chain hoist (5) to lifting sling (1). Guide cylinder head (2) straight up while soldier B lifts it.

Soldier B 3. Work chain hoist (5) to lift front cylinder head assembly (2) off engine (6) . Set front cylinder head assembly down on workbench.

Soldiers A and B 4. Do steps 1, 2, and 3 again for rear cylinder head assembly (7).

GO TO FRAME 3

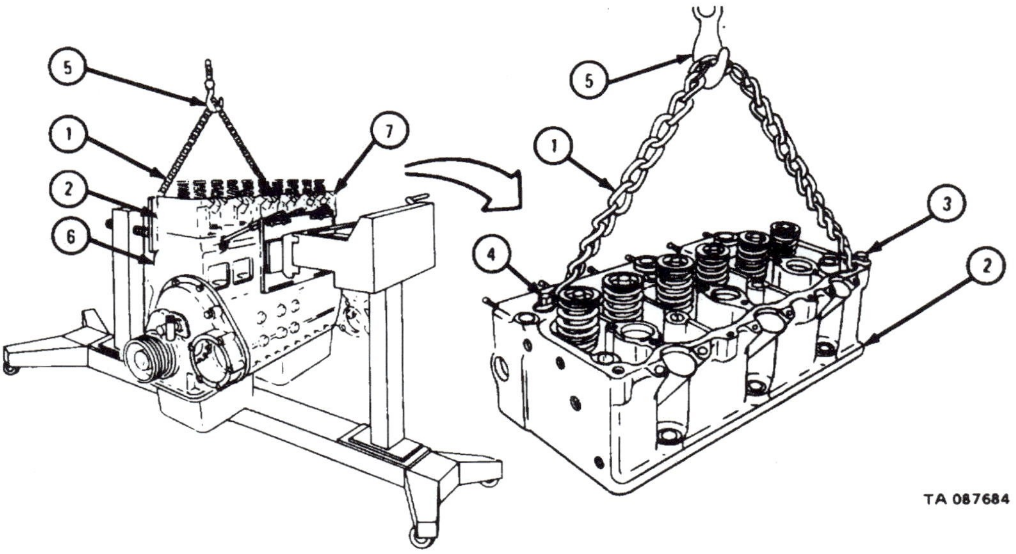

FRAME 3

1. Take off and throw away front cylinder head gasket (1) and three fire rings (2).
2. Take off and throw away rear cylinder head gasket (3) and three fire rings (4).

END OF TASK

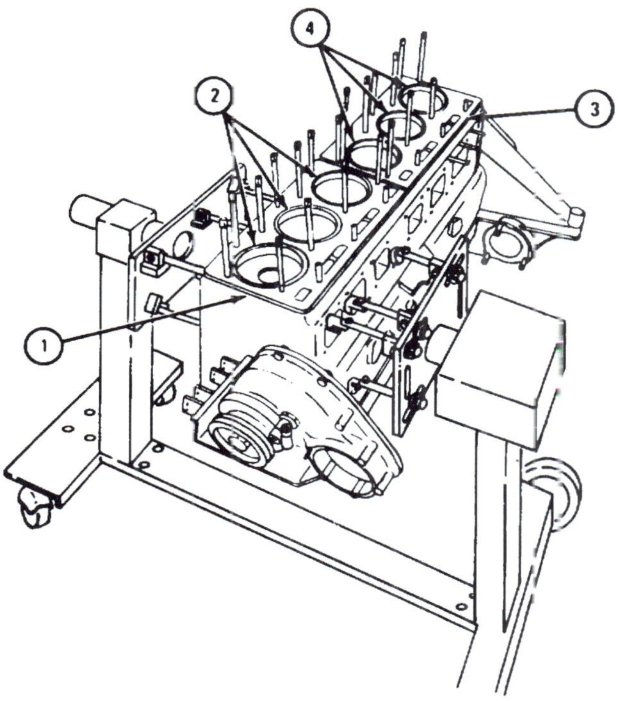

TA 113614

3-35. CLUTCH ASSEMBLY (ALL ENGINES EXCEPT LDS-465-2).

a. Engines LD-465-1, LD-465-1C, and LDT-465-1C .

FRAME 1

1. Using prybar, press in four clutch release levers (1), one at a time, and put a clutch release lever spacer block (2) between each clutch release lever and pressure plate assembly (3) as shown .

GO TO FRAME 2

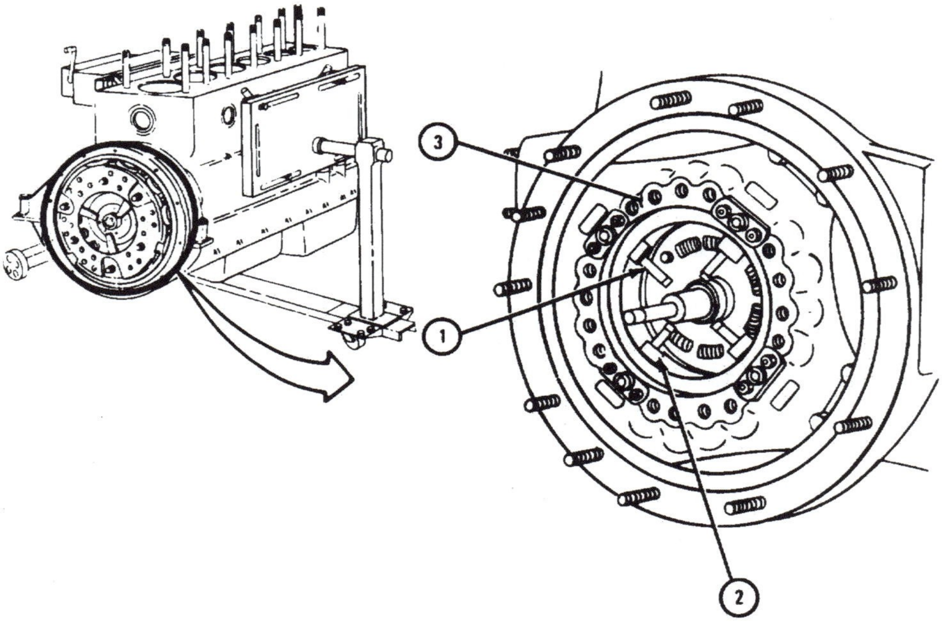

TA 087685

FRAME 2

WARNING

pressure plate assembly (1) is heavy. It must be held firmly up and against flywheel (2) when mounting screws are taken out. If it is not held up, it will fall and cause injury to personnel .

Clutch disk (4) is heavy enough to cause injury if dropped. It is held in place by pressure plate assembly (1). Do not let it fall out and drop when pressure plate assembly is taken out or it may cause injury to personnel.

NOTE

Clutch disk (4) may have been taken out before engine was shipped .

Soldier A	1.	Hold pressure plate assembly (1) firmly up and against flywheel (2) so it does not fall when soldier B takes out mounting screws (3).
Soldier B	2.	Take out eight screws and lockwashers (3) .
Soldiers A and B	3.	Take out pressure plate assembly (1) and clutch disk (4).

END OF TASK

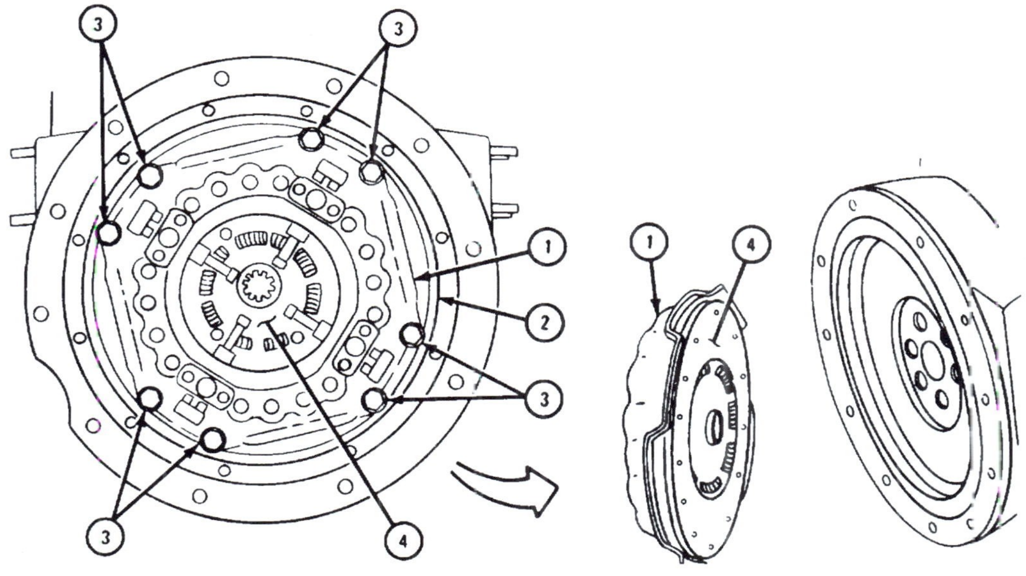

b. Engines LDS-465-1 and LDS-465-1A.

FRAME 1

CAUTION

Do not make a punch mark anywhere on pressure plate assembly (1) except where pressure plate assembly is flush with rim of flywheel (2). Pressure plate assembly could be bent .

1. Make a punch mark on pressure plate assembly (1) and rim of flywheel (2) as shown. Both marks should be next to each other so pressure plate assembly and flywheel can be alined during engine assembly.

NOTE

There are many holes in pressure plate assembly (1). Only three holes have a threaded sleeve behind them for retaining screws (3) . The location of these holes is different on different pressure plate assemblies. The right holes may be hard to find because threaded sleeves are inside springs (4) .

2. Using strong light, look through big hole in center of pressure plate assembly (1). Find the three holes with threaded sleeves behind them.

3. Put 3/8 - 16 x 2 1/4-inch retaining screw (3) and 1 1/4-inch outside diameter, 1 /8-inch thick flat washer (5) into the three holes found in step 2. Tighten three retaining screws (3) evenly to take spring pressure off clutch disk on flywheel (2).

GO TO FRAME 2

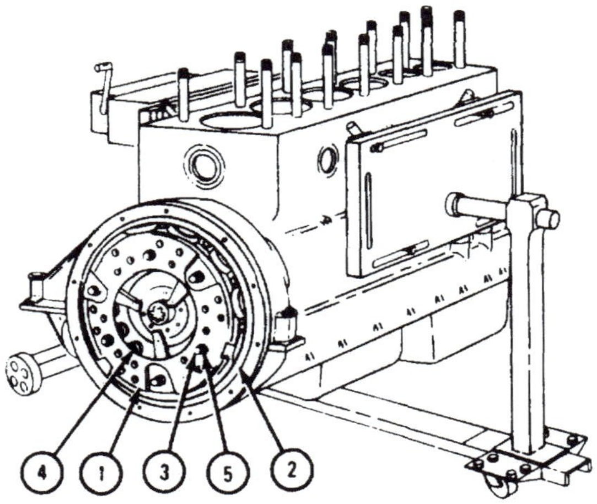

TA 087691

FRAME 2

WARNING

Pressure plate assembly (1) is very heavy. It must be held firmly up and against flywheel (2) when mounting screws are taken out. If it is not held up, it will fall and cause injury.

Clutch disk (3) is heavy enough to cause injury if dropped. It is held in place by pressure plate assembly (1). Do not let it fail out and drop when pressure plate assembly is taken off or it may cause injury to personnel.

NOTE

Clutch disk (3) may have been taken out before engine was shipped.

Soldier A	1.	Hold pressure plate assembly (1) firmly up and against flywheel (2) so it does not fall when soldier B takes out mounting screws (4).
Soldier B	2.	Take out 12 screws and lockwashers (4).
Soldiers A and B	3.	Take out pressure plate assembly (1) and take out clutch disk (3).

END OF TASK

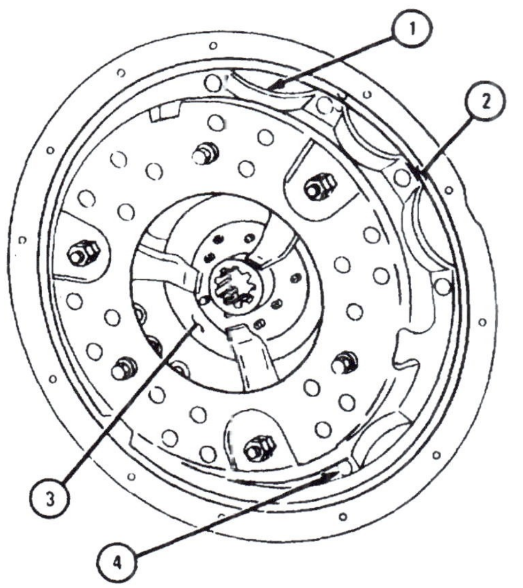

3-36. FLYWHEEL .

a. Engines LD-465-1, LD-465-1C, and LDT-465-1C .

NOTE

One threaded hole in crankshaft flange (1) and one mounting hole in flywheel (2) are offset about 1/16 inch. Because of this, flywheel (2) can be put back on crankshaft flange (1) in only one position.

1. Scribe a line on crankshaft flange (1) and edge of flywheel (2), so flywheel can be put back in same position on crankshaft flange.

Soldier A 2 . Using engine barring tool, hold crankshaft damper and pulley (3) so it cannot turn to the right.

Soldier B 3 . Take top lockbolt (4) out of flywheel.

GO TO FRAME 2

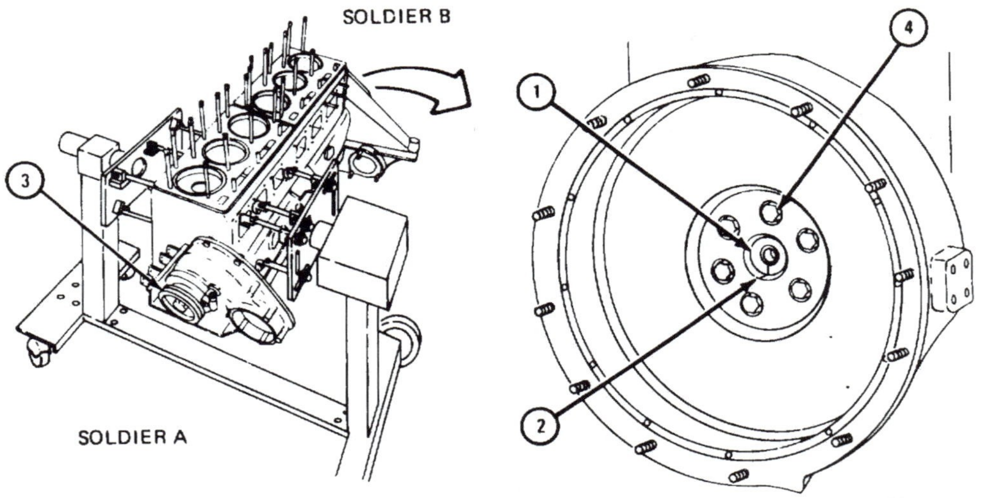

TA 087689

FRAME 2

NOTE

Tell machine shop to make pilot stud (1). See figure 3-3.

1. Put pilot stud (1) in hole of flywheel (2) as shown.

Soldier A 2. Using engine barring tool, hold crankshaft damper and pulley (3) so it cannot turn to the right.

Soldier B 3. Hold flywheel (2) against crankshaft so it will not fall when lockbolts (4) are taken out. While holding flywheel against crankshaft, take out five lockbolts (4).

Soldiers A and B 4. Hold up flywheel (2) and slide it back clear of pilot stud (1) and take it off. Set flywheel down on wooden blocks.

Soldier A 5. Take out pilot stud (1).

END OF TASK

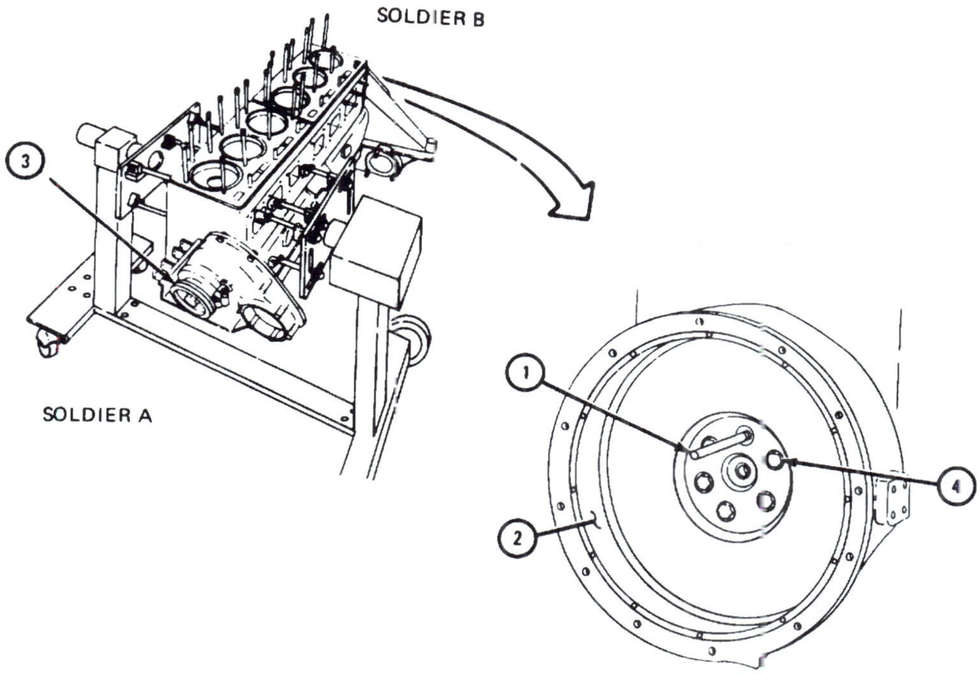

TA 087690

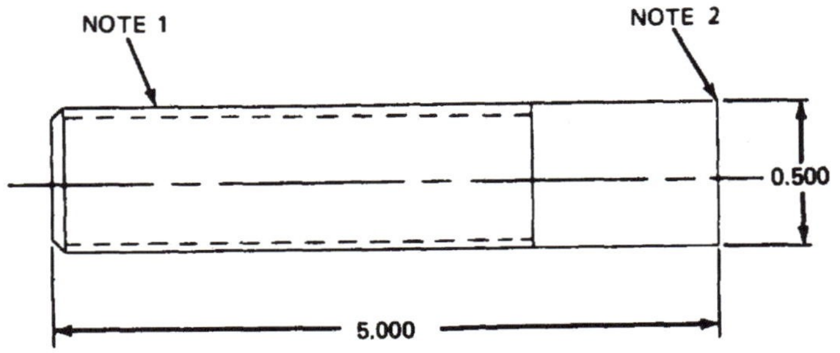

NOTES:

1. MATERIAL - USE 5-INCH LONG BOLT X 1/2-INCH DIAMETER X 20 THREADS PER INCH
2. CUT OFF HEAD IN AREA (2). GRIND OFF SHARP EDGES

TA 121162

Figure 3-3. Pilot Stud Fabrication Instructions

b. Engines LDS-465-1 and LDS-465-1A.

FRAME 1

NOTE

One threaded hole in crankshaft flange and one mounting hole in flywheel (1) are offset about 1/6 inch. Because of this, flywheel can be put back on crankshaft flange in only one position.

Soldier A 1. Using engine barring tool, hold crankshaft damper and pulley (2) so it cannot turn to the right.

Soldier B 2. Take top lockbolt (3) out of flywheel (1).

GO TO FRAME 2

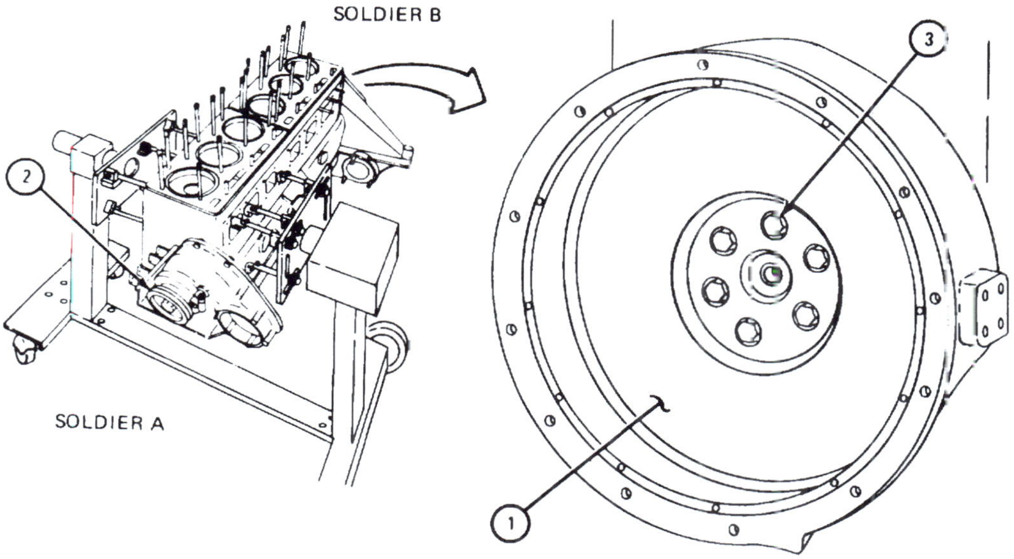

1A 087693

FRAME 2

NOTE

Tell machine shop to make pilot stud (1). See figure 3-3.

1. Put pilot stud (1) in mounting hole in flywheel (2) as shown.
2. Make a scribe mark (3) on flywheel (2) next to pilot stud (1) as shown.

GO TO FRAME 3

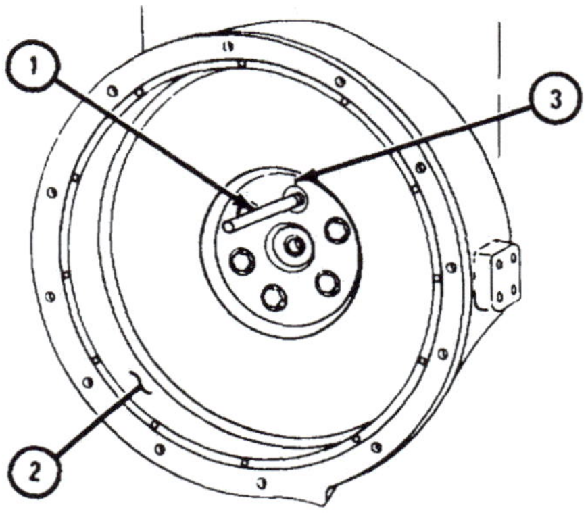

TA 086356

FRAME 3

Soldier A 1.	Using engine barring tool, hold crankshaft damper and pulley (1) so it cannot turn to the right.
Soldier B 2.	Hold flywheel (2) against crankshaft so it will not fall when lockbolts (3) are taken out. While holding flywheel against crankshaft, take out five lockbolts (3).
Soldiers A and B 3.	Hold up flywheel (2) and slide it back clear of pilot stud (4) and off. Set flywheel down on wood blocks.
Soldier A 4.	Mark crankshaft flange (5) around hole for pilot stud (4) with lacquer. Take out pilot stud (4).

END OF TASK

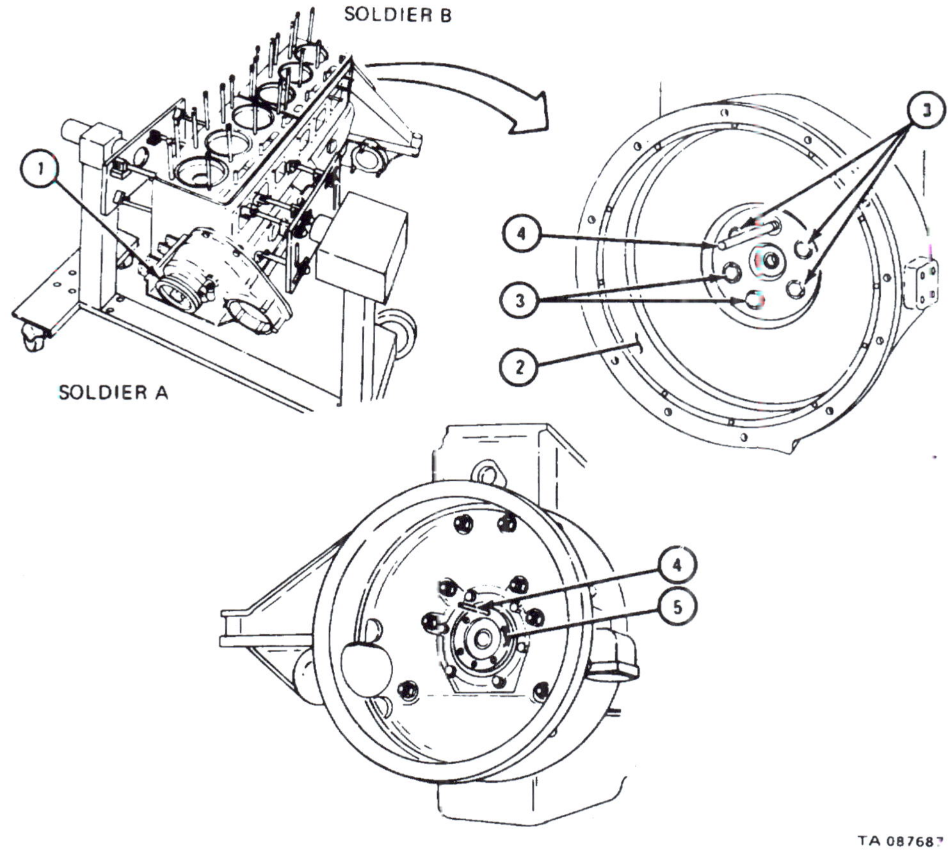

c. Engine LDS-465-2 .

FRAME 1

NOTE

One threaded hole in crankshaft flange and one mounting hole in flywheel (1) are offset about 1/16 inch. Because of this, flywheel (1) and flexible transmission plates (2) can be put back on crankshaft flange in only one position .

Soldier A 1 . Using engine barring tool, hold crankshaft damper and pulley (3) so it cannot turn to the right.

Soldier B 2 . Take out whichever lockbolt (4) is now near top of flywheel (1).

GO TO FRAME 2

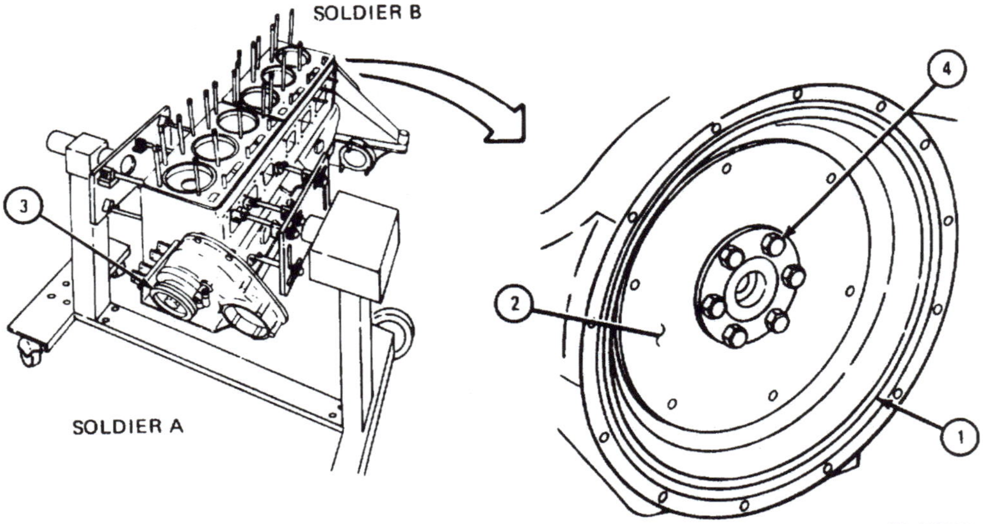

TA 087694

FRAME 2

NOTE

Tell machine shop to make pilot stud (1). See figure 3-3.

1. Put pilot stud (1) through hole in flywheel (2) and flexible transmission plates (3) and tighten as shown.
2. Make a scribe mark (4) on transmission drive ring (5) next to pilot stud (1) as shown. Make a second pair of scribe marks (6) near edge of outer flexible drive plate (3) and on inner lip of flywheel (2) opposite pilot bolt (1) as shown.

GO TO FRAME 3

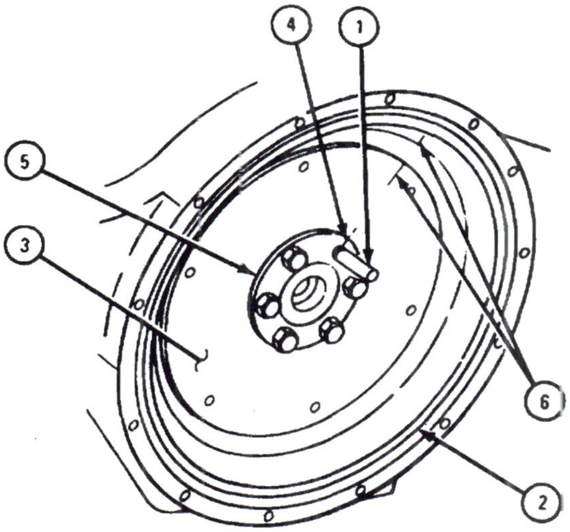

TA 087695

FRAME 3

Soldier A 1. Using engine barring tool, hold crankshaft damper and pulley (1) so it cannot turn to the right.

Soldier B 2. Hold flywheel (2) and flexible drive plates (3) against crankshaft so they will not fall when lockbolts (4) are taken out. While holding flywheel (2) and flexible drive plates (3) against crankshaft, take out five lockbolts (4).

3. Take out transmission drive ring (5) and outer flexible drive plate (3).

Soldier A 4. Using tool tag and wire, tag outer flexible drive plate (3) with the number 4 so it can be put back in the same place.

G0 TO FRAME 4

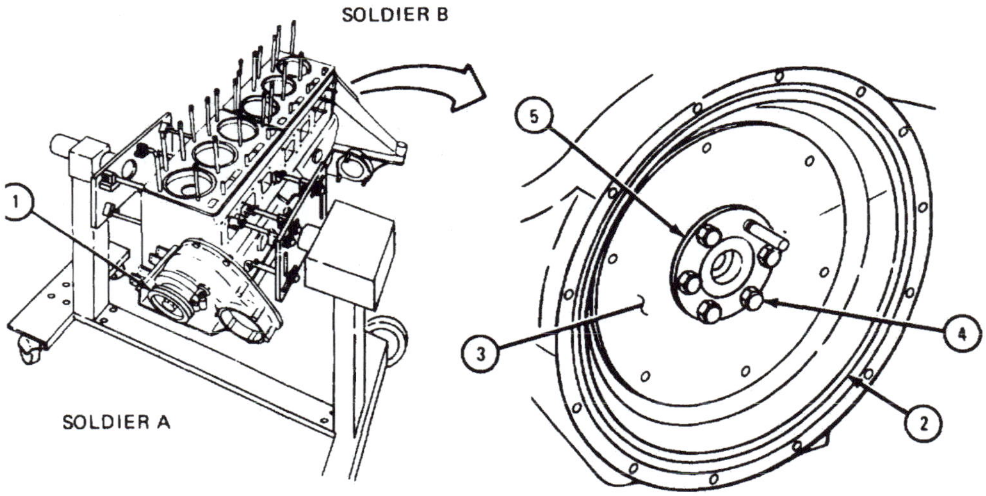

TA 087696

FRAME 4

NOTE

Flexible drive plate (1) was already taken out.

		1.	Make a scribe mark (2) on flexible drive plate (3) alined with scribe mark (4) on inner lip of flywheel (5) as shown.
Soldier	B	2.	Take off flexible drive plate (3).
Soldier	A	3.	Using tool tag and wire, tag flexible drive plate (3) with the number 3 so it can be put back in the same place.
		4.	Make a scribe mark (6) on flexible drive plate (7) alined with scribe mark (4) on inner lip of flywheel (5) as shown.
Soldier	B	5.	Take off flexible drive plate (7).
Soldier	A	6.	Using tool tag and wire, tag flexible drive plate (7) with the number 2 so it can be put back in the same place.

GO TO FRAME 5

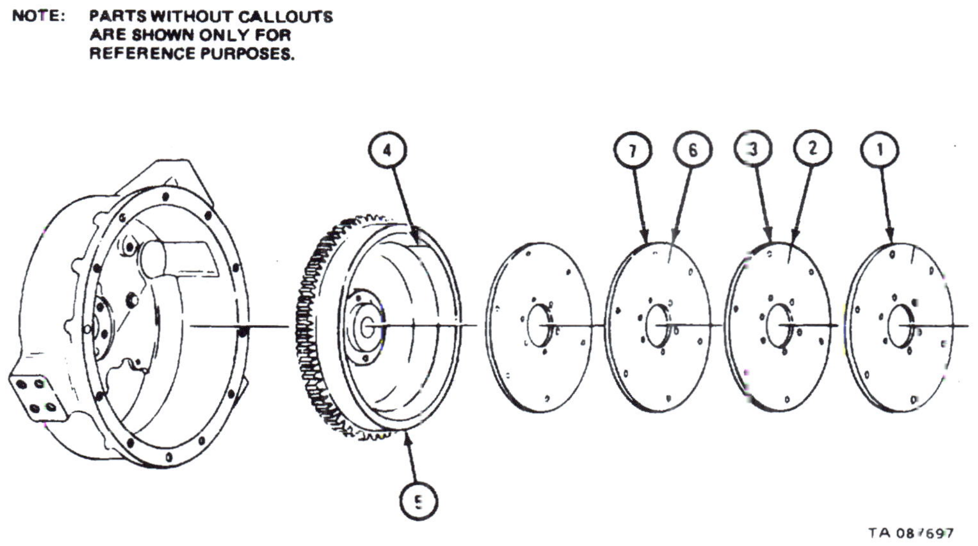

FRAME 5

1. Make a scribe mark (1) on last flexible drive plate (2) alined with scribe mark (3) on inner lip of flywheel (4) as shown.

Soldier B 2. Take off last flexible drive plate (2).

Soldier A 3. Using tool tag and wire, tag last flexible drive plate (2) with the number 1 so it can be put back in the same place.

GO TO FRAME 6

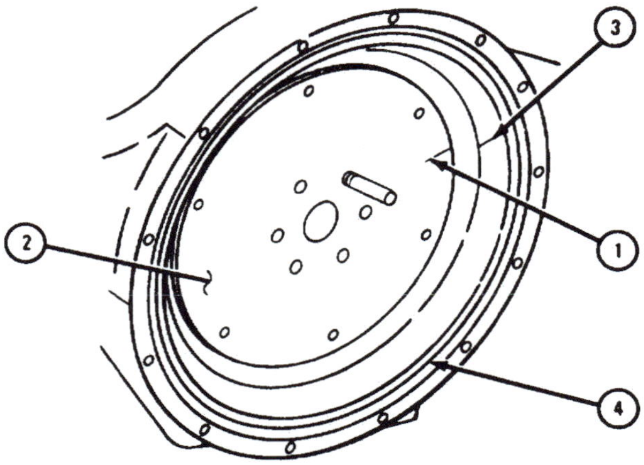

TA 087688

FRAME 6

Soldiers A and B 1. Hold up flywheel (1) and slide it back clear of pilot stud (2) and take it off. Set flywheel down on wood blocks.

Soldier A 2. Using lacquer, mark crankshaft flange (3) around hole for pilot stud (2).

3. Take out pilot stud (2).

END OF TASK

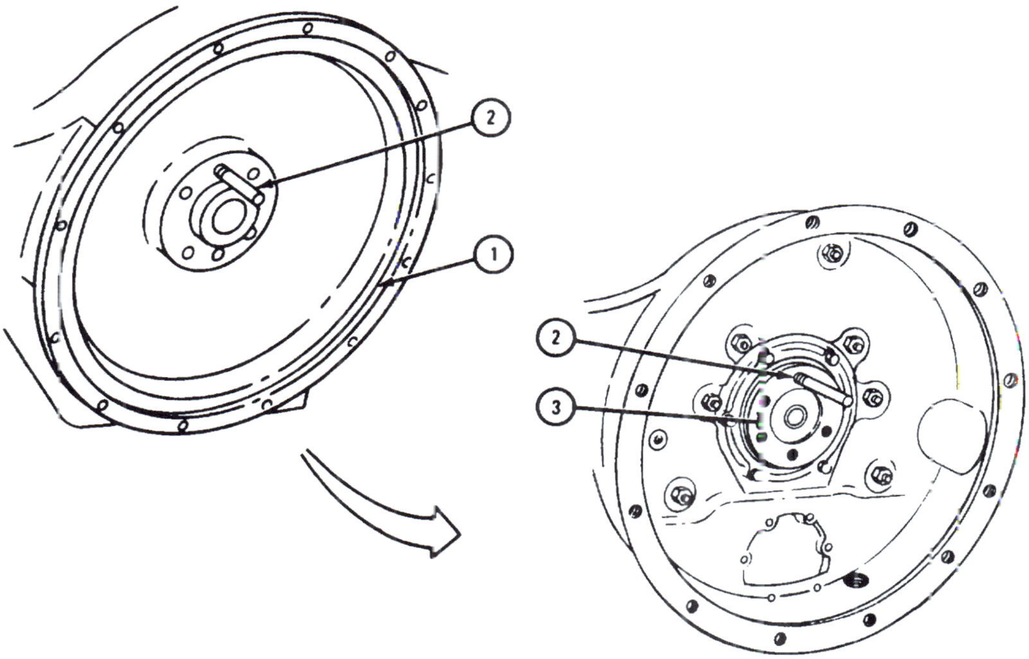

TA 113615

3-37. FLYWHEEL HOUSING.

FRAME 1

NOTE

Flywheel housing (3) shown is for engine LDS-465-1A. This task is the same for all other engines.

1. Take off two nuts (1) and two 1/8-inch thick washers (2).

WARNING

Flywheel housing (3) is heavy. Be sure to hold it firmly up and against engine when taking off mounting nuts (4). It could fall off mounting studs (5) and cause injury to personnel.

Soldier A 2. Hold flywheel housing (3) firmly against engine so it does not slip off mounting studs (5) and fall.

Soldier B 3. Take off six nuts (4) and six 1/16-inch washers (6).

Soldiers A and B 4. Take off flywheel housing and gasket (3). Throw away gasket.

END OF TASK

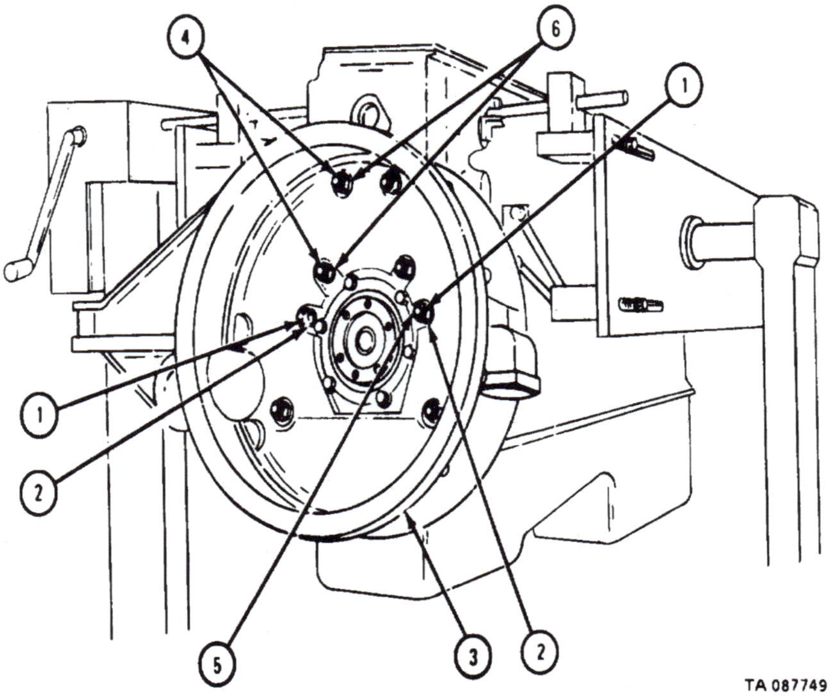

3-38. CRANKSHAFT REAR OIL SEAL AND HOUSING.

FRAME 1

NOTE

Some engines do not have oil seal housing gasket (1). On these engines, flywheel housing gasket (2) cannot be taken off until crankshaft rear oil seal housing (3) is taken off.

1. When working on an engine with an oil seal housing gasket (1), take off flywheel housing gasket (2) and throw it away.

NOTE

Most engines do not have two lockplates (4).

2. Bend back tabs on lockplates (4). Take out six screws and lockwashers (5) and two lockplates (4). Take off crankshaft rear oil seal housing (3) and oil seal (6).

3. Using a small punch, drive oil seal (6) out of rear oil seal housing (3). Throw away oil seal.

4. When working on an engine with an oil seal housing gasket (1), take it off and throw it away.

5. If working on an engine which does not have an oil seal housing gasket (1), take off flywheel housing gasket (2) and throw it away.

END OF TASK

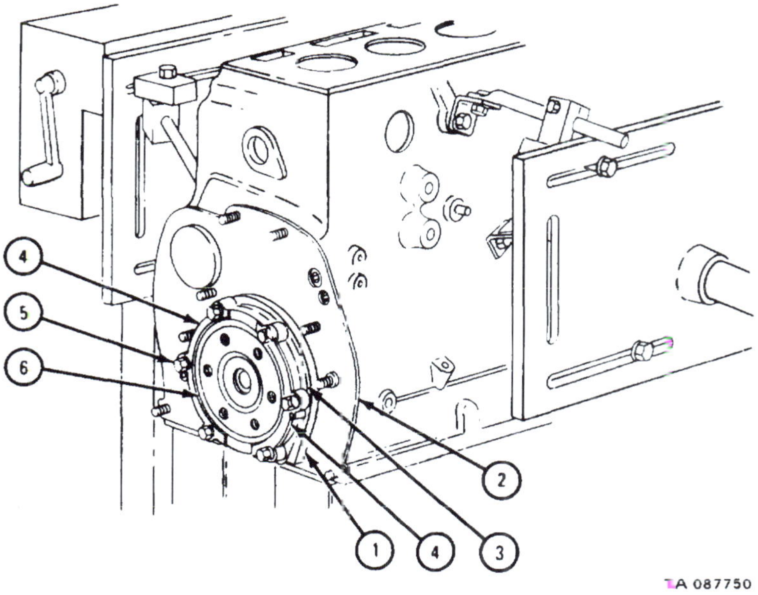

3-39. OIL PAN.

a. Engines LD-465-1, LD-465-1A, and LDT-465-1C.

FRAME 1

	1.	Work engine stand to turn engine upside down so oil pan (1) faces up as shown.
	2.	Takeout 30 screws and lockwashers (2).
Soldiers A and B	3.	Take off oil pan (1).
Soldier A	4.	Take off and throw away oil pan gasket (3).

END OF TASK

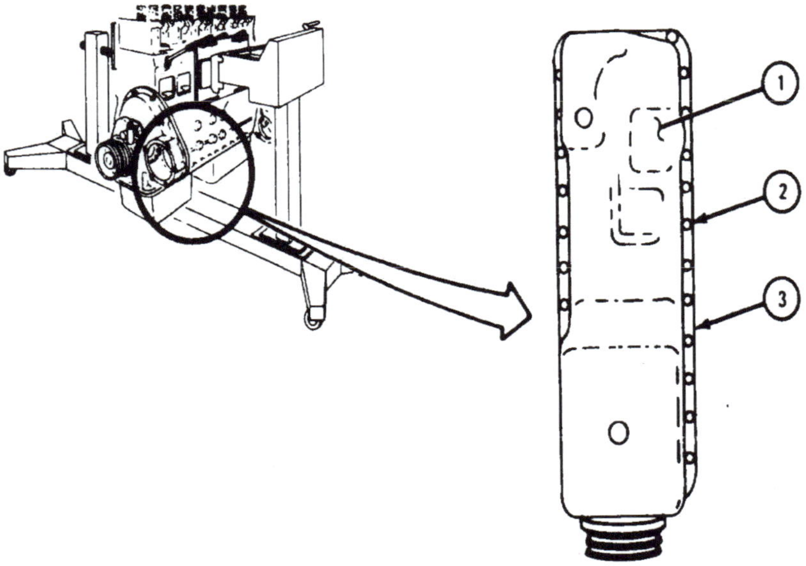

b. Engine LDS-465-1 .

FRAME 1

1. Work engine stand to turn engine upside down so oil pan (1 faces up as shown.

NOTE

If engine had side-mounted flame heater system, three screws and lockwashers (2) were taken out with flame heater fuel pump and fuel filter bracket. Screw and lockwasher (3) were taken out with flame heater fuel supply and fuel return tube clamp .

2. Take out three screws and lockwashers (2) and screw and lockwasher (3) .

NOTE

Some LDS-465-1 engines do not have fuel injection pump oil drain hose clamp bracket (4) .

3. Take out screw and lockwasher (5) and fuel injection pump oil drain hose clamp bracket (4) .

4. Take out 25 screws and lockwashers (6) .

Soldiers A and B 5. Take off oil pan (1) .

Soldier A 6. Take off and throw away oil pan gasket (7) .

END OF TASK

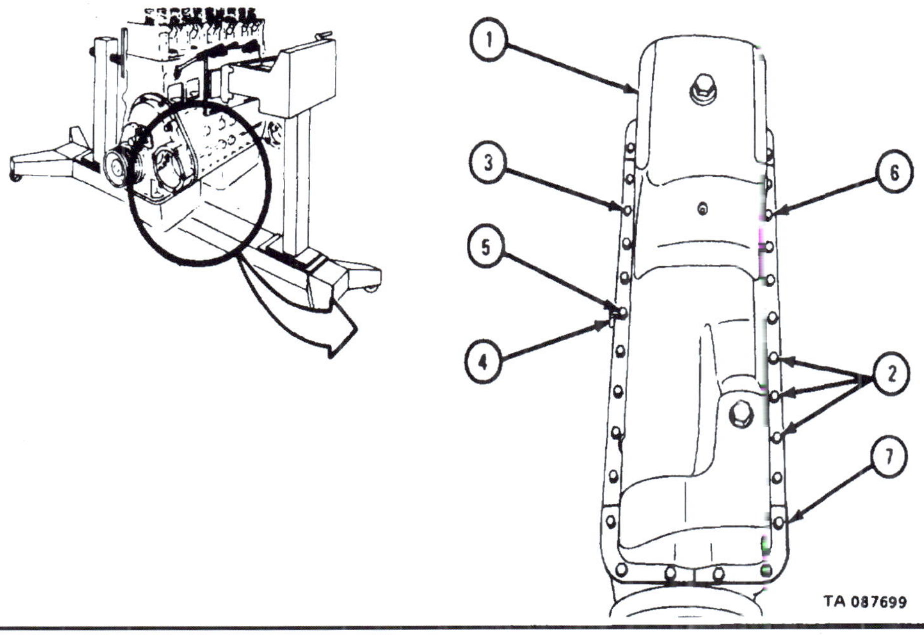

c. Engine LDS-465-1A.

FRAME 1

1. Work engine stand to turn engine upside down so oil pan (1) faces up as shown.

2. Take out screw and internal tooth lockwasher (2). Take off engine ground lead (3).

NOTE

Some LDS-465-1A engines do not have fuel filter drain tube bracket (5).

3. Take out screw and lockwasher (4). Take off fuel filter drain tube bracket (5).

4. Take out 28 screws and lockwashers (6).

Soldiers A and B 5. Take off oil pan (1).

Soldier A 6. Take off and throw away oil pan gasket (7).

END OF TASK

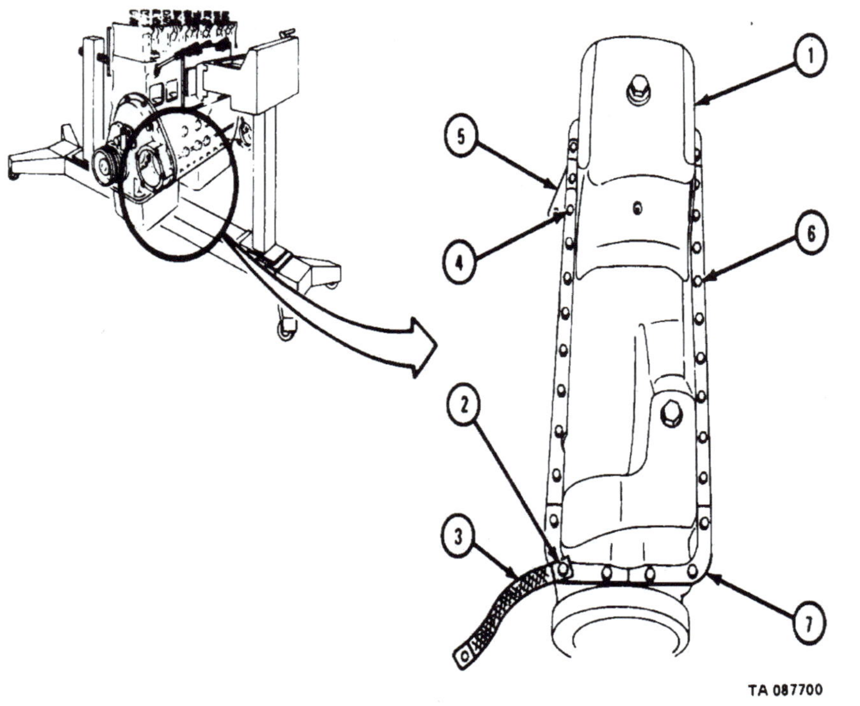

d. Engine LDS-465-2.

FRAME 1

	1. Work engine stand to turn engine upside down so oil pan (1) faces up as shown.
	2. Take out two screws and lockwashers (2). Take out throttle return spring bracket (3).
	3. Take out 28 screws and lockwashers (4).
Soldiers A and B	4. Take off oil pan (1).
Soldier A	5. Take off and throw away oil pan gasket (5).

END OF TASK

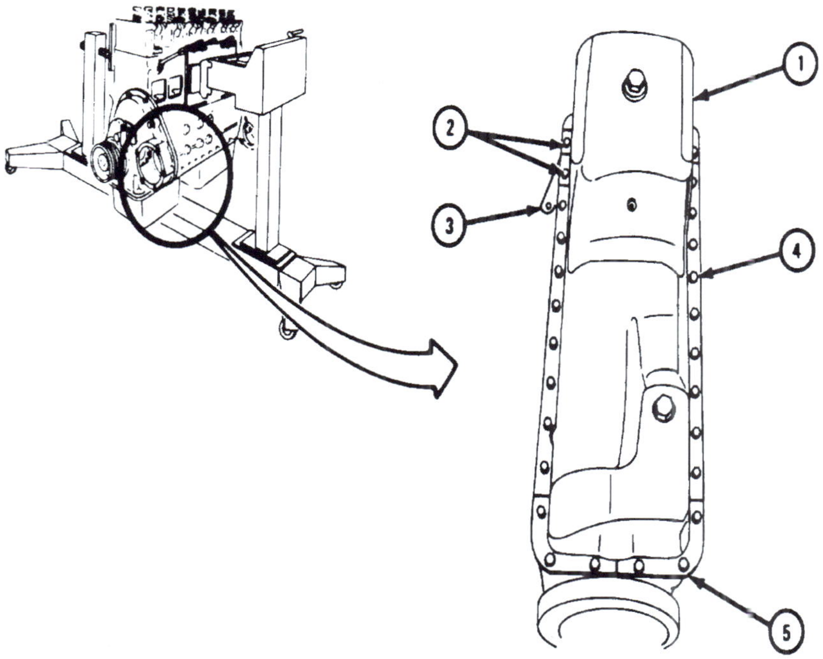

3-40. OIL PUMP TUBES.

a. Engines LD-465-1, LD-465-1C, and LDT-465-1C.

FRAME 1

1. Take out screw, lockwasher and nut (1).
2. Take out two screws and lockwashers (2). Take out pressure oil pump pickup tube (3) with tube clamp (4). Take off pressure oil pump pickup tube gasket (5) and throw it away.

GO TO FRAME 2

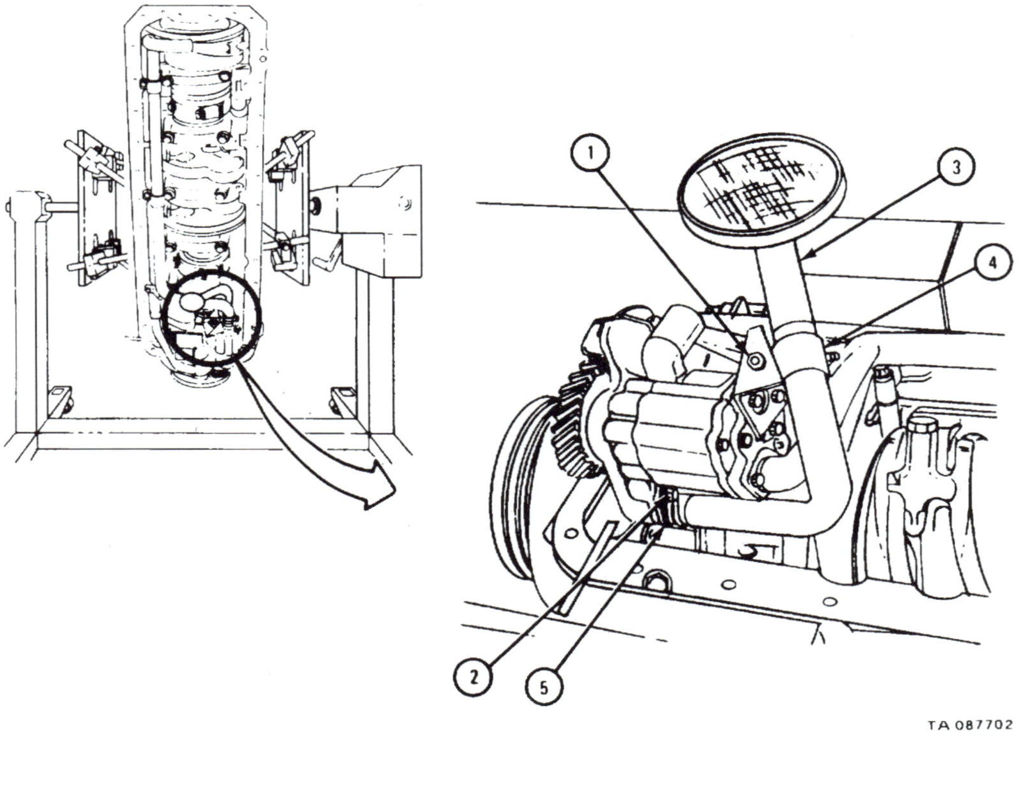

FRAME 2

1. Cut and takeout safetywire (1) . Takeout two drilled-head bolts (2) .
2. Take out two screws and lockwires (3) .
3. Take out pressure oil pump outlet tube (4), outlet tube flange (5), and pressure oil pump outlet tube flange O-ring (6). Throw away O-ring .
4. Take off outlet tube gasket (7) and throw it away .

GO TO FRAME 3

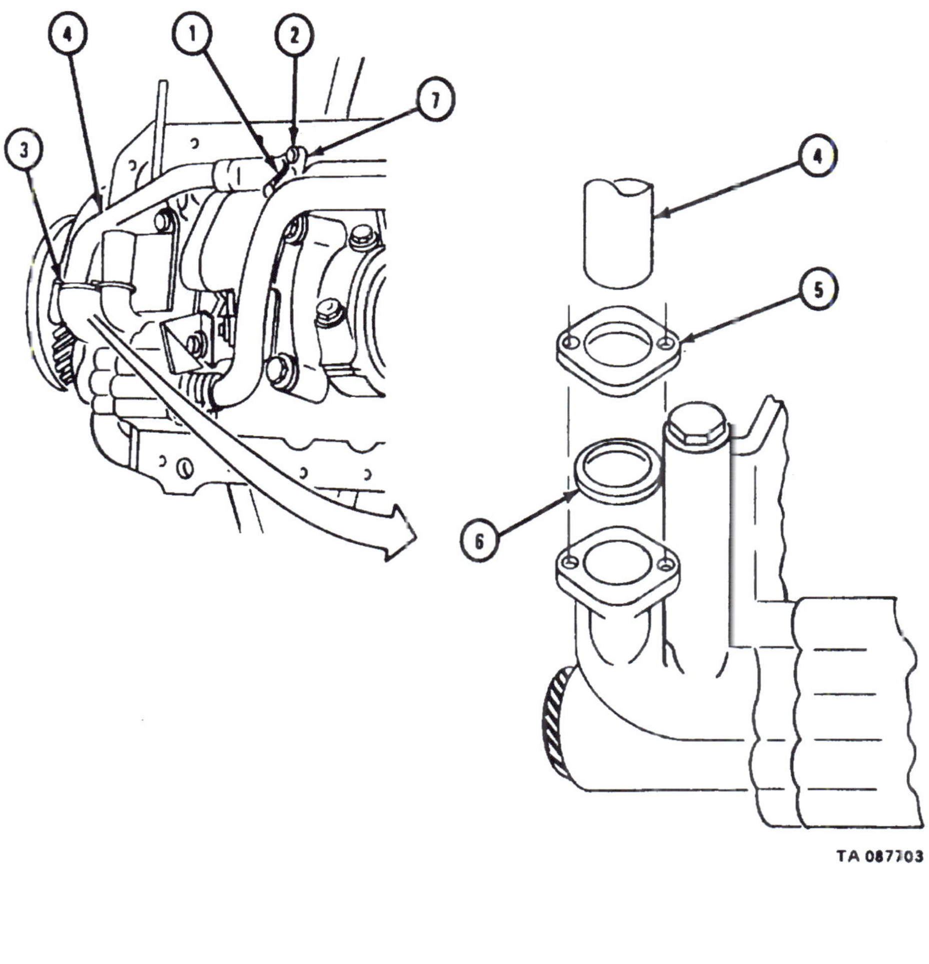

FRAME 3

1. Take out two screws and lockwashers (1) .
2. Take out two screws and lockwashers (2) .
3. Take out scavenge oil pump inlet tube (3) with two tube clamps (4) .
4. Take off inlet tube gasket (5) and throw it away .

END OF TASK

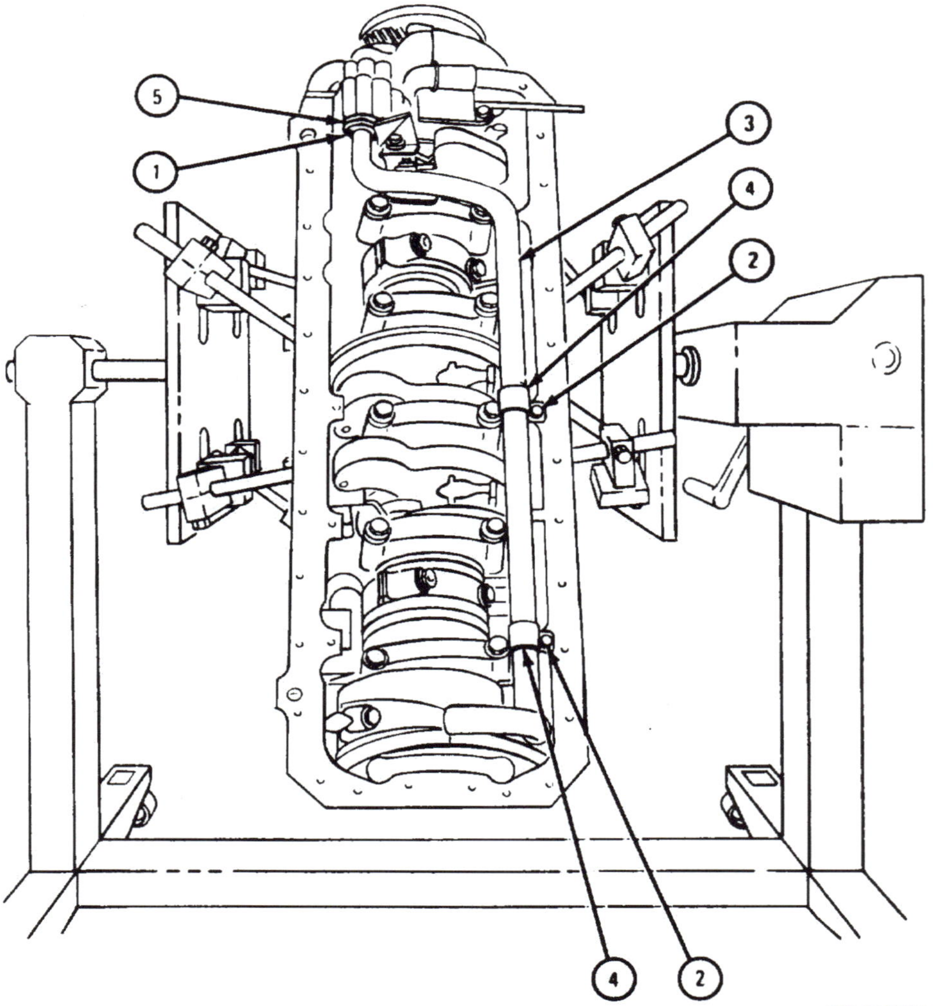

b. Engines LDS-465-1, LD-465-1A, and LDS-465-2.

NOTE

Only the pressure oil pump outlet tube is taken out before the oil pump is removed from these engines.

FRAME 1

1. Cut and take out safetywire (1). Take out two drilled-head bolts (2).
2. Take out two screws and lockwashers (3).
3. Take out pressure oil pump outlet tube (4), outlet tube flange (5), and pressure oil pump outlet tube flange O-ring (6). Throw away O-ring.
4. Take off outlet tube gasket (7) and throw it away.

END OF TASK

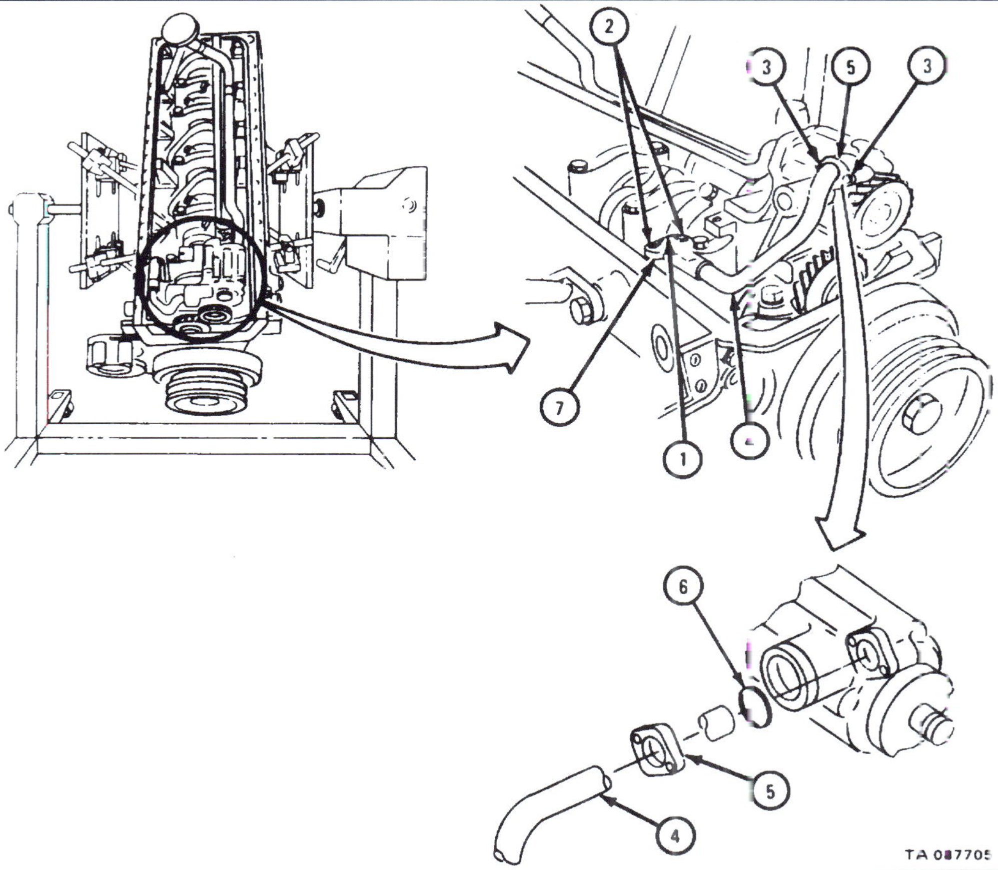

3-41. OIL PUMP ASSEMBLY.

a. Oil Pump Gear Backlash Check.

FRAME 1

1. Mount dial indicator on timing gear cover (1). Point of dial indicator should rest against side of a tooth on oil pump drive gear (2).

NOTE

When measuring backlash, make sure that oil pump idler gear (3) does not turn. If gear turns, backlash readings will be wrong.

2. Turn oil pump drive gear (2) as far as you can in one direction. Set dial indicator to read 0.
3. Hold oil pump idler gear (3) so it cannot turn. Turn oil pump drive gear (2) as much as you can in the other direction. Read backlash between oil pump drive gear (2) and oil pump idler gear (3).
4. If backlash is not between 0.006 and 0.014 inch, mark oil pump drive gear (2) and oil pump idler gear (3) for replacement.

END OF TASK

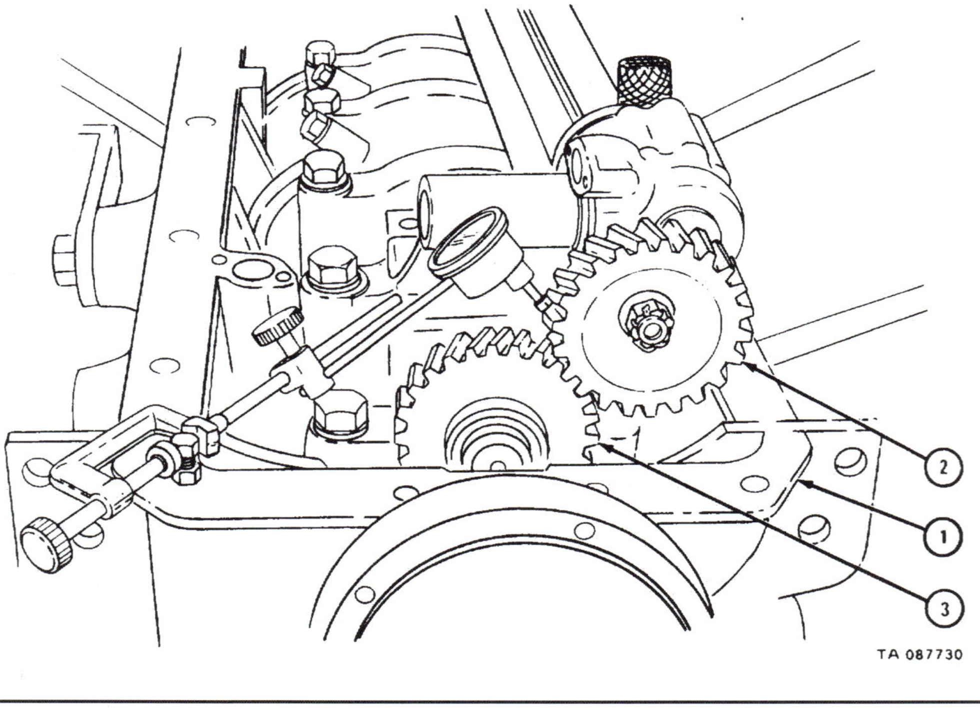

Removal .

(1) Engines LD-465-1, LD-465-1C, and LDT-465-1C .

FRAME 1

CAUTION

Oil pump assembly (2) must be pulled straight out of engine. If it is not pulled straight out, small dowel pins behind oil pump may be broken.

1. Take out three lockbolts (1) and take out oil pump assembly (2) .

END OF TASK

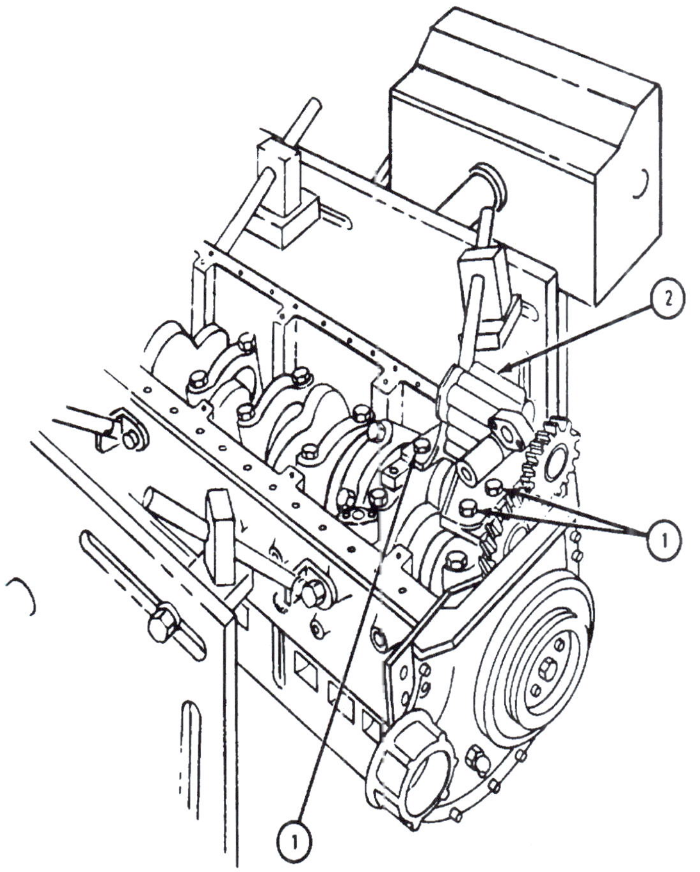

(2) Engines LDS-465-1, LDS-465-1A, and LDS-465-2.

FRAME 1

1. Take out screw and lockwasher (1).
2. Takeout screw and lockwasher (2).

GO TO FRAME 2

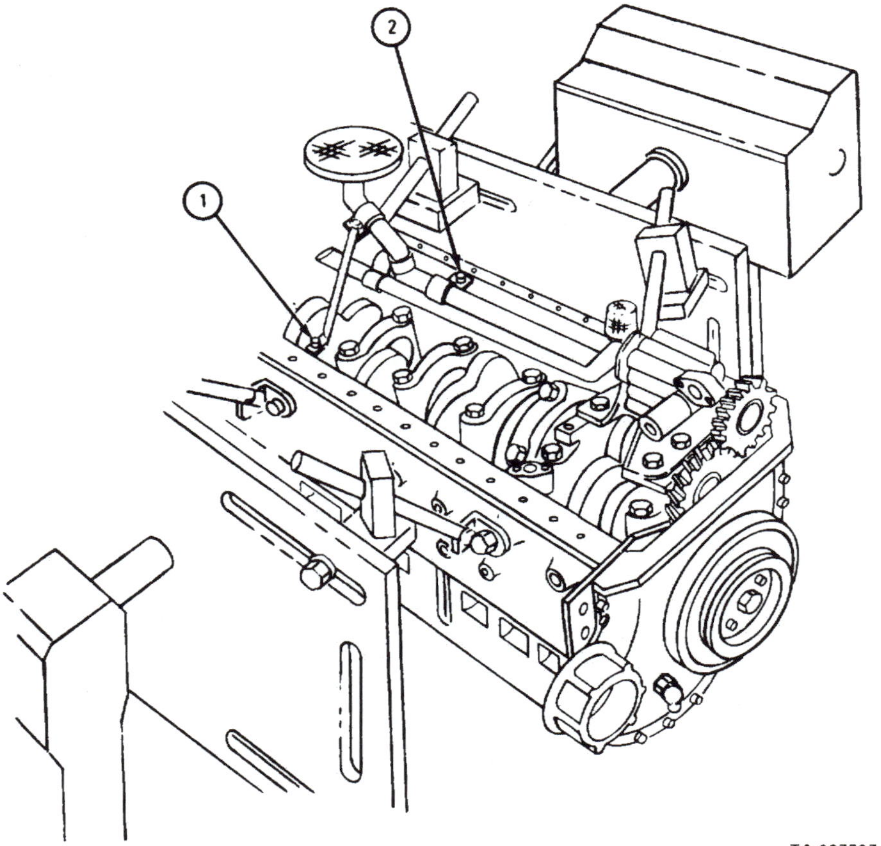

TA 087707

FRAME 2

CAUTION

Oil pump assembly (2) must be pulled straight out of engine. If it is not pulled straight out, small dowel pins behind oil pump assembly may be broken.

1. Take out three lockbolts (1) and take out oil pump assembly (2) with tubes (3, 4, and 5).

GO TO FRAME 3

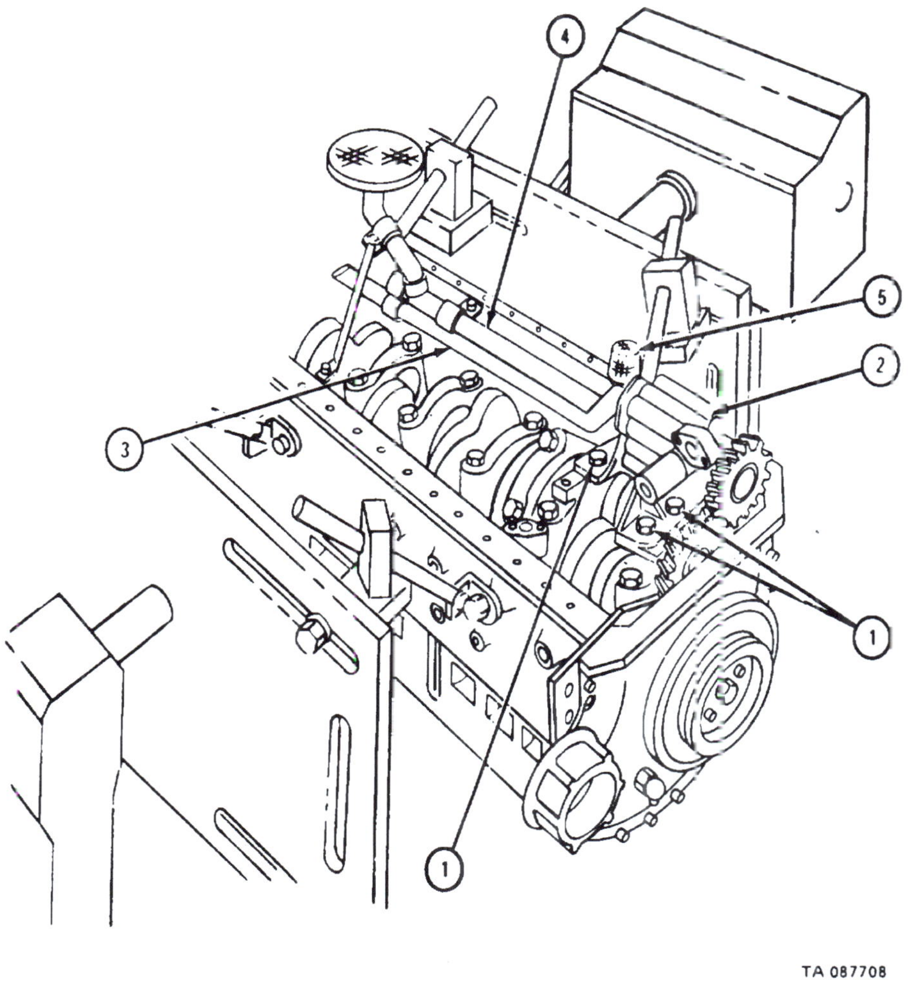

FRAME 3

1. Take out two screws and lockwashers (1) and take off scavenge oil pump inlet tube (2). Take off and throw away scavenge oil pump inlet tube gasket (3).

2. Take out four screws and lockwashers (4). Take off pressure oil pump pickup tube (5) and scavenge oil pump outlet tube (6) together. Take off and throw away two gaskets (7).

GO TO FRAME 4

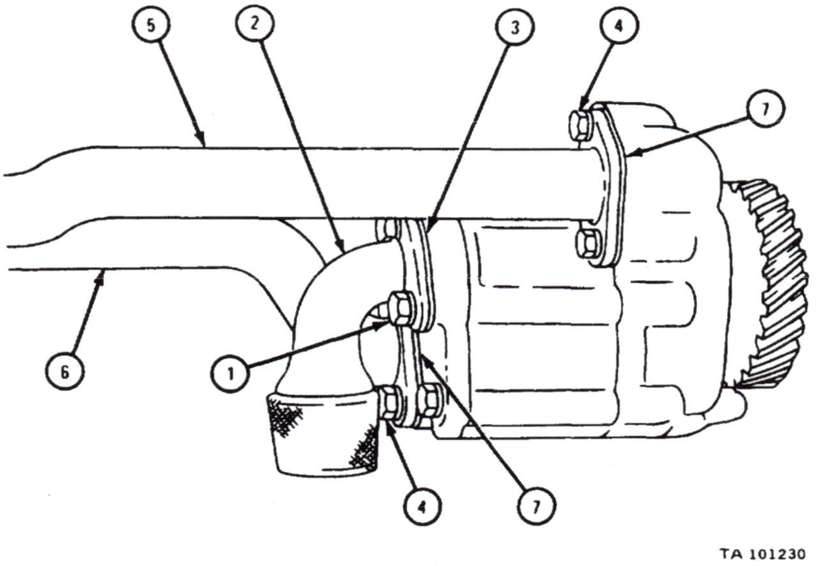

FRAME 4

1. Take out nut (1), lockwasher (2), and screw (3) and take off pressure oil pump pickup tube brace (4).

2. Take out nut (5), lockwasher (6), and screw (7) and take scavenge oil pump outlet tube (8) apart from pressure oil pump pickup tube (9).

END OF TASK

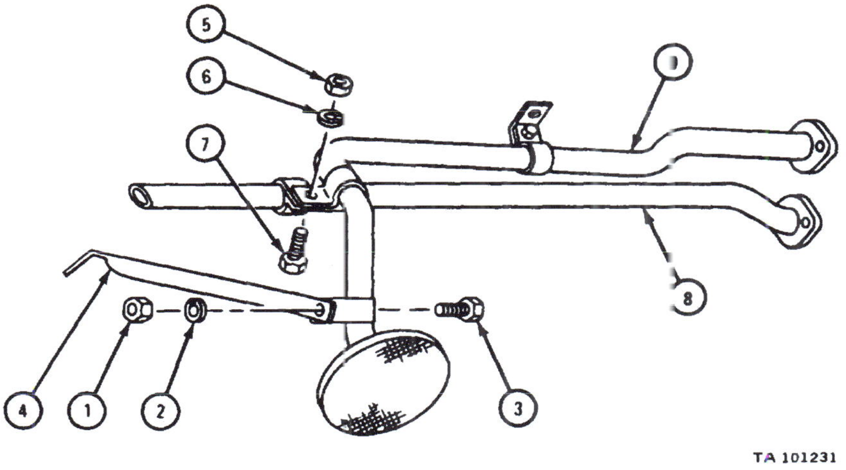

3-42. PISTON AND CONNECTING ROD ASSEMBLIES.

CAUTION

It is easy to damage the equipment if you do not know what you are doing. Do not try to do this task unless you are experienced at it, or you have an experienced person with you.

FRAME 1

1. Turn engine overhaul stand so that crankshaft damper and pulley (1) are facing up.
2. Using engine barring tool, turn crankshaft damper and pulley (1) until cylinder number one piston and connecting rod assembly (2) and cylinder number six piston and connecting rod assembly (3) are at bottom of stroke.

NOTE

Cylinder sleeves (4, 5, and 6) and piston and connecting rod assemblies (2, 3, and 7) are called by numbers one to six counting from front to rear of engine (8).

3. Using cylinder ridge reamer, ream out ridge at top of number one cylinder sleeve (4) and number six cylinder sleeve (5).

GO TO FRAME 2

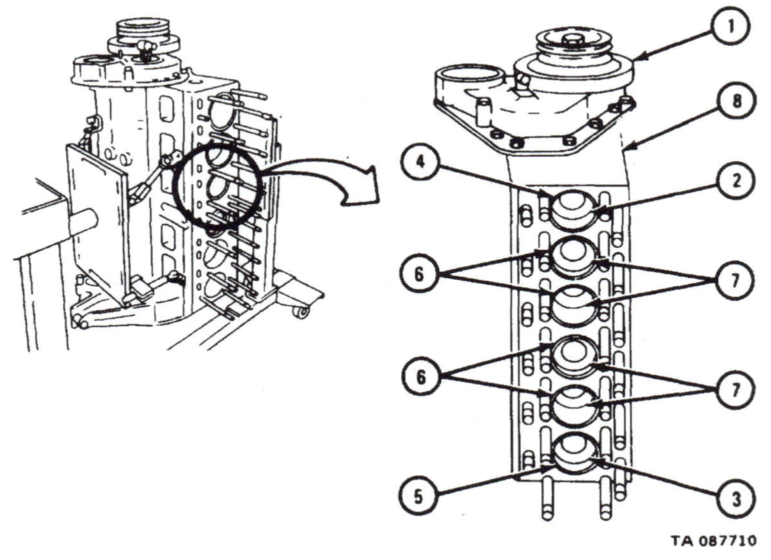

FRAME 2

NOTE

Cylinders and piston and connecting rod assemblies are called by numbers one to six counting from front to rear of engine.

1. Turn crankshaft damper and pulley (1) until two bolts (2) holding cylinder number one connecting rod bearing cap (3) can be easily reached, as shown.
2. Take out two bolts (2) and take out cylinder number one connecting rod bearing cap (3).
3. Take off cylinder number one connecting rod bearing half (4). Tag it with number of piston and connecting rod assembly it was taken from, followed by letter "C" for cap.
4. Take out two bolts (5) and take out cylinder number six connecting rod bearing cap (6).
5. Do step 3 again for number six connecting rod bearing half (4).

GO TO FRAME 3

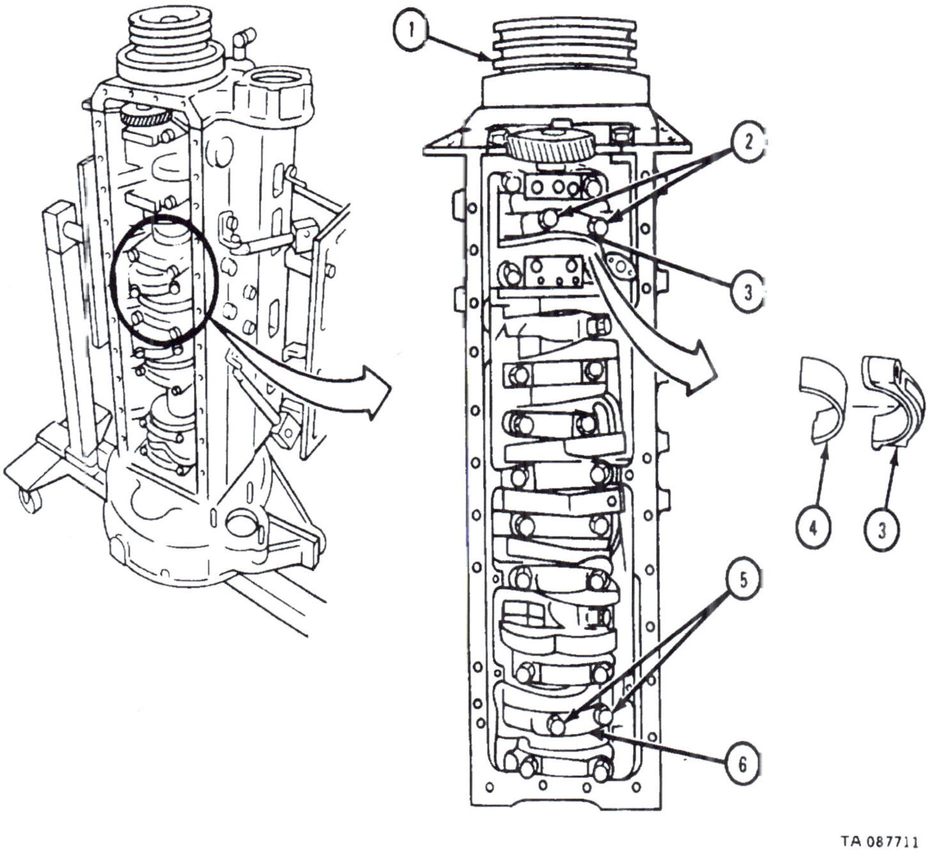

TA 087711

FRAME 3

1. Turn crankshaft damper and pulley (1) about 1/4 turn to the left to leave spaces between cylinder number one connecting rod (2) and connecting rod journal (3), and between cylinder number six connecting rod (4) and connecting rod journal (5).

GO TO FRAME 4

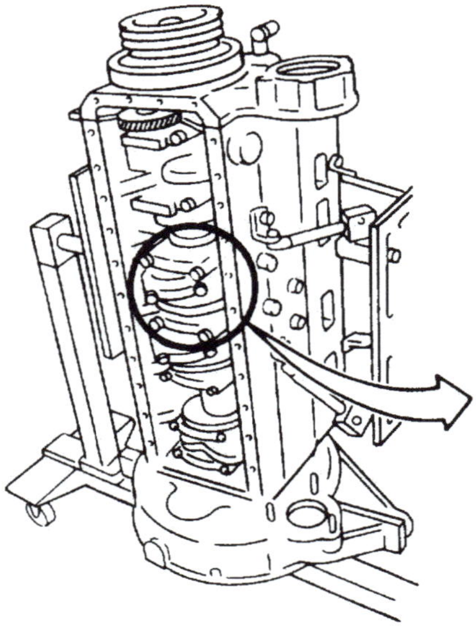

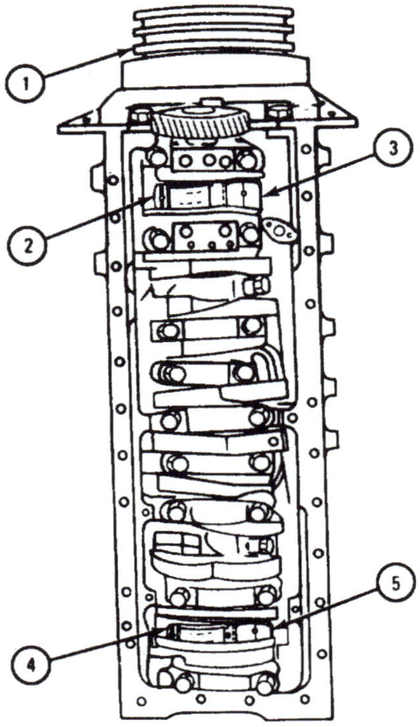

FRAME 4

CAUTION

Be careful not to scratch inside surface of cylinder sleeves (1) when taking out piston and connecting rod assemblies (2 and 3).

Be careful not to scratch or dent connecting rod upper bearing halves (4) when pushing out piston and connecting rod assemblies (2 and 3). Push only on bottom of connecting rod (5), not on bearing halves.

NOTE

Cylinder sleeves (1) may come out with piston and connecting rod assemblies (2 and 3). If cylinder sleeve comes out, tag it so it can be put back in the same place.

Soldier A	1.	Watch top of cylinder number one piston and connecting rod assembly (2). Tell soldier B when piston rings (6) are clear of cylinder sleeve (1).
Soldier B	2.	Using wood or nylon dowel and hammer, gently tap bottom of cylinder number one connecting rod (5). Drive piston and connecting rod assembly (2) out of engine until soldier A tells you piston rings (6) are clear of cylinder sleeve (1).
Soldier A	3.	Pull piston and connecting rod assembly (2) out of engine.
Soldiers A and B	4.	Do steps 1 through 3 again for cylinder number six piston and connecting rod assembly (3).

GO TO FRAME 5

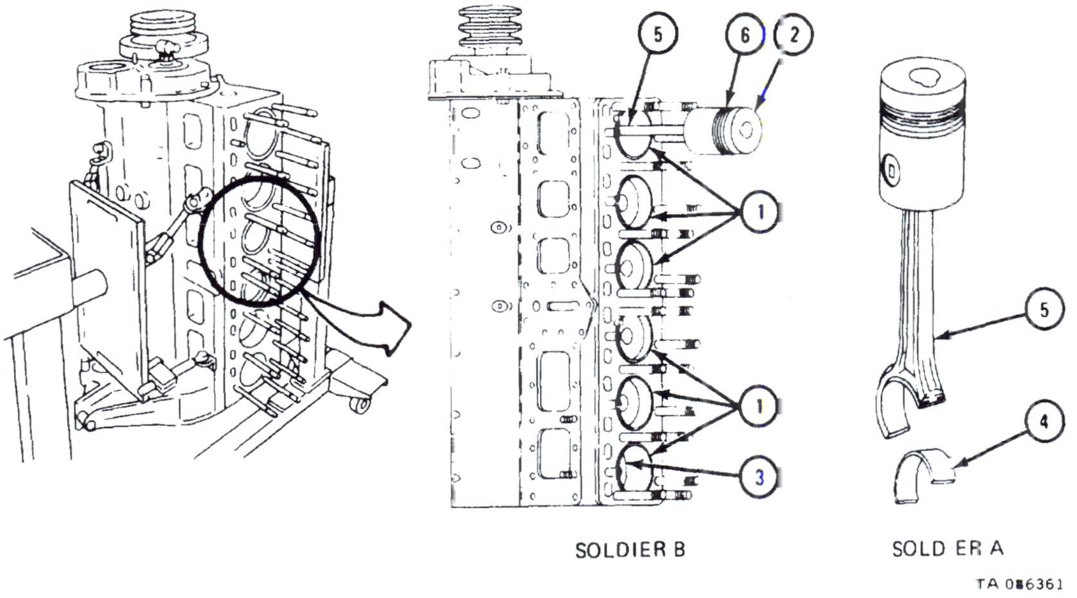

FRAME 5

1. Takeout connecting rod upper bearing half (1).
2. Tag connecting rod upper bearing half (1) with number of cylinder it came from and letter "R" for connecting rod.

GO TO FRAME 6

NOTE: PARTS WITHOUT CALLOUTS ARE SHOWN ONLY FOR REFERENCE PURPOSES.

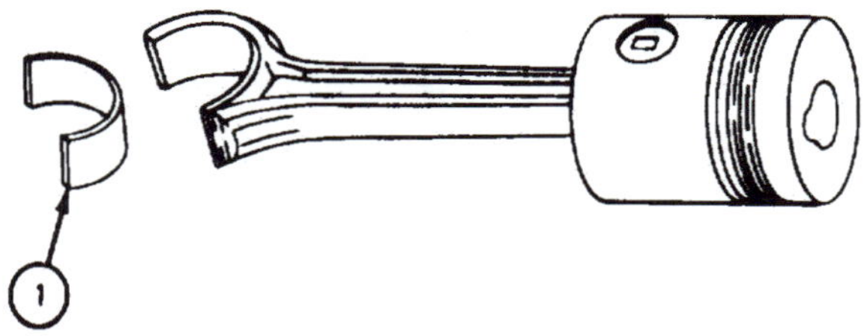

TA 087713

FRAME 6

NOTE

Piston and connecting rod assemblies (1 through 4) are taken out in pairs.

1. Do frames 1 through 5 again for cylinder number two piston and connecting rod assembly (1) and cylinder number five piston and connecting rod assembly (2).
2. Do frames 1 through 5 again for cylinder number three piston and connecting rod assembly (3) and cylinder number four piston and connecting rod assembly (4).

END OF TASK

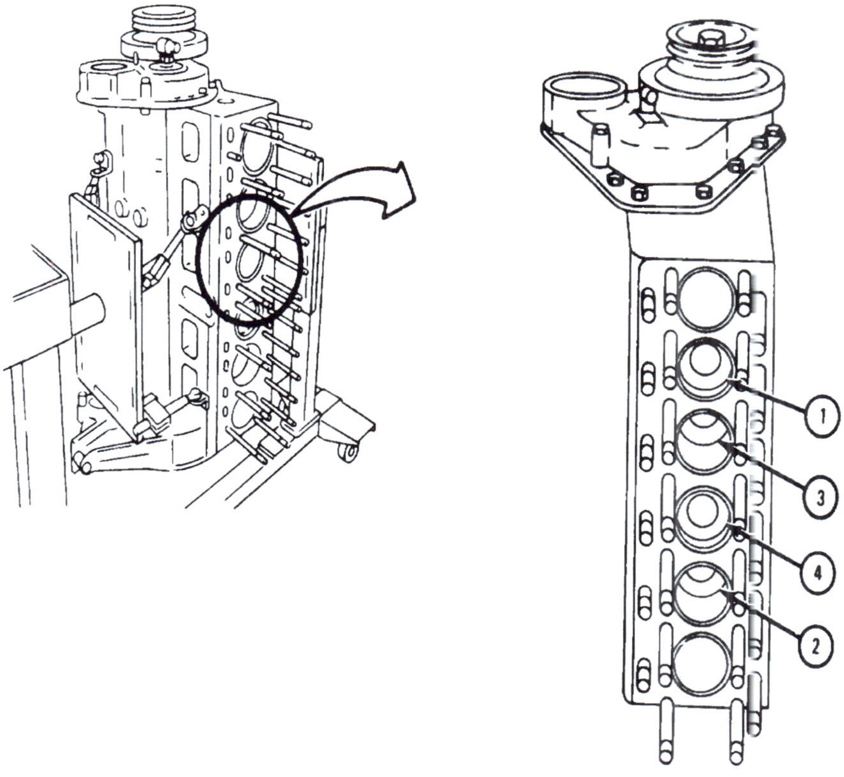

TA 113613

3-43. CRANKSHAFT DAMPER AND PULLEY ASSEMBLY.

FRAME 1

Soldier A 1. Put two bolts (1) in crankshaft flange (2) as shown.

2. Wedge a length of steel bar stock (3) between two bolts (1) so that other end of bar rests on floor. This will keep crankshaft from turning when crankshaft damper and pulley retaining bolt is taken out. Hold steel bar in this position while soldier B takes out crankshaft damper and pulley retaining bolt.

GO TO FRAME 2

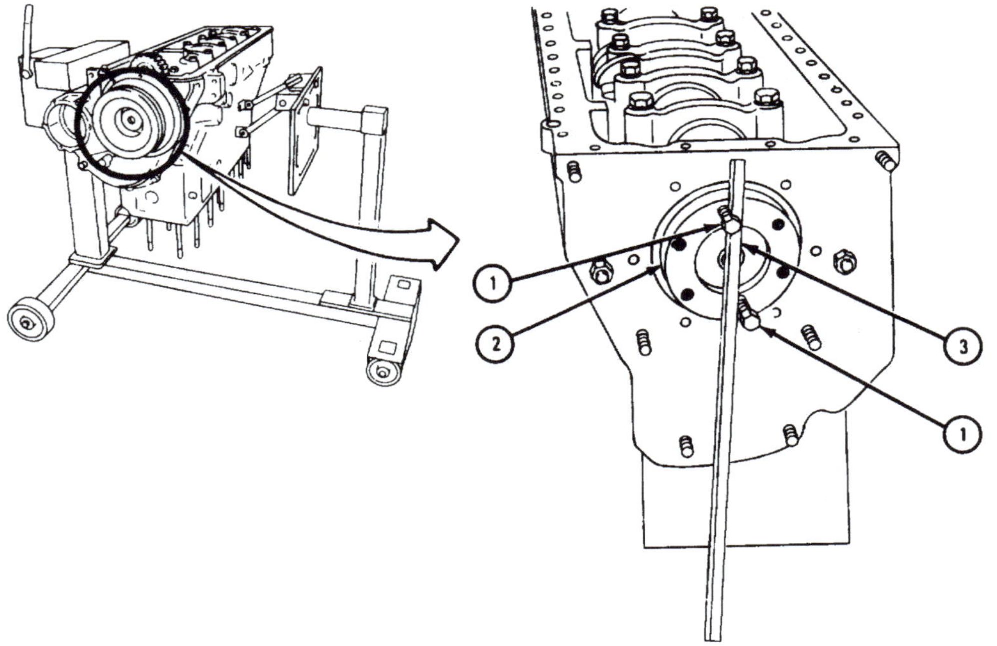

FRAME 2

Soldier B 1. Takeout two screws and lockwashers (1) and take off lockplate (2).

2. Using impact wrench, take out crankshaft damper and puller retaining bolt (3).

3. Take off crankshaft damper and pulley retaining washer (4).

WARNING

Crankshaft damper and pulley assembly (5) is heavy. Dropping it will cause damage to equipment and injury to personnel.

4. If working on any engine except LDS - 465-2, take off crankshaft damper and pulley assembly (5) using puller kit and plug.

5. If working on engine LDS-465-2, take off crankshaft damper and pulley assembly (5) using mechanical puller.

END OF TASK

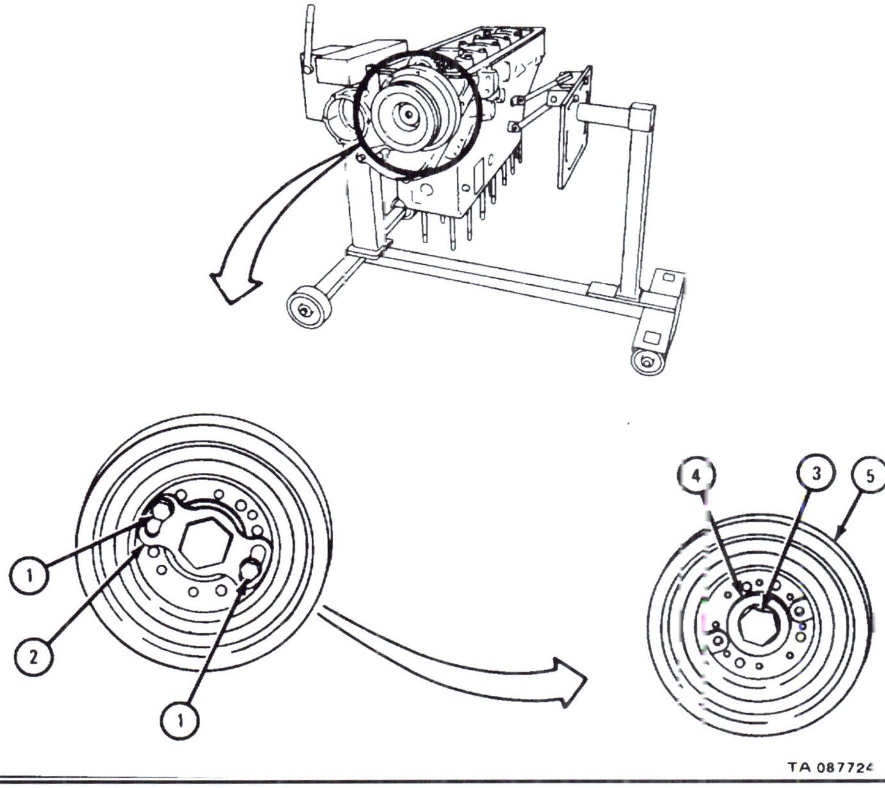

3-44. TACHOMETER DRIVE ADAPTER.

a. All Engines Except LDS-465-2.

FRAME 1

1. Unscrew and takeout tachometer drive adapter (1).
2. Pull out tachometer drive shaft (2).
3. Unscrew and take out tachometer takeoff adapter (3). Take off and throw away tachometer takeoff adapter gasket (4).

END OF TASK

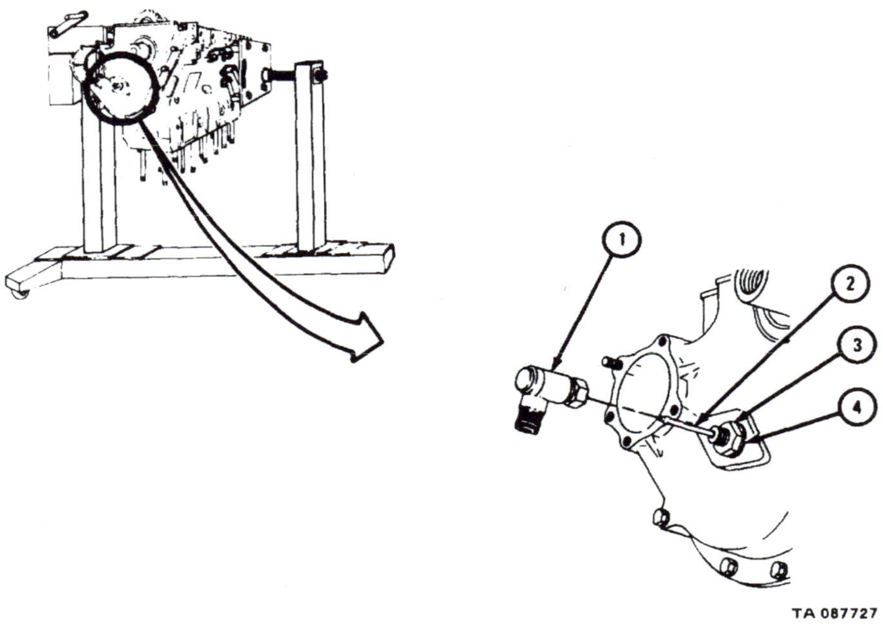

b. Engine LDS-465-2.

FRAME 1

CAUTION

Be careful not to bend drive shaft (1) when taking out tachometer drive adapter and shaft assembly (2).

1. Unscrew and takeout tachometer drive adapter and shaft assembly (2). Take off and throw away gasket (3).

2. Unscrew and takeout tachometer takeoff adapter (4). Takeoff and throw away tachometer takeoff adapter gasket (5).

END OF TASK

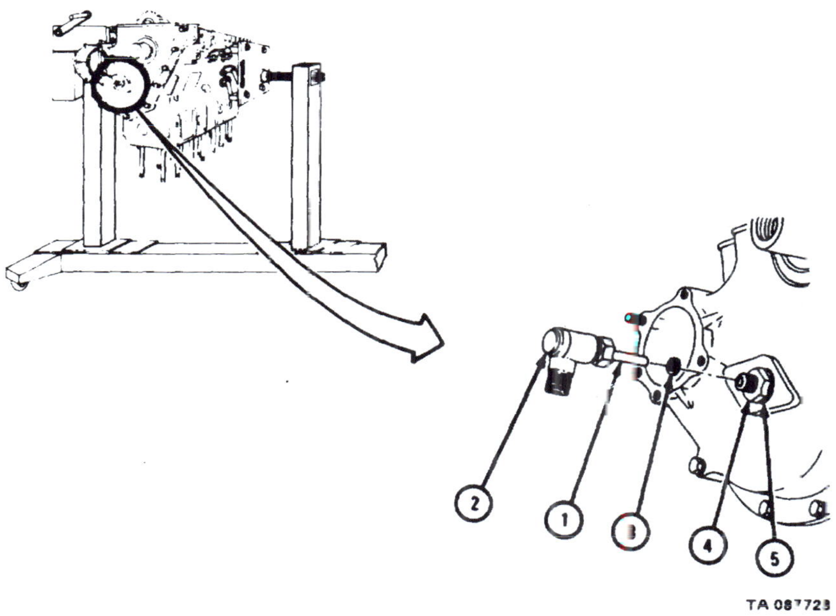

3-45. TIMING GEAR COVER ASSEMBLY.

FRAME 1

1. Take out six screws and lockwashers (1).
2. Take off nut, lockwasher, and screw (2).
3. Take off nut, lockwasher, and screw (3).
4. Take off nut and lockwasher (4), and take off timing gear cover (5). Take off and throw away timing gear cover gasket (6).

END OF TASK

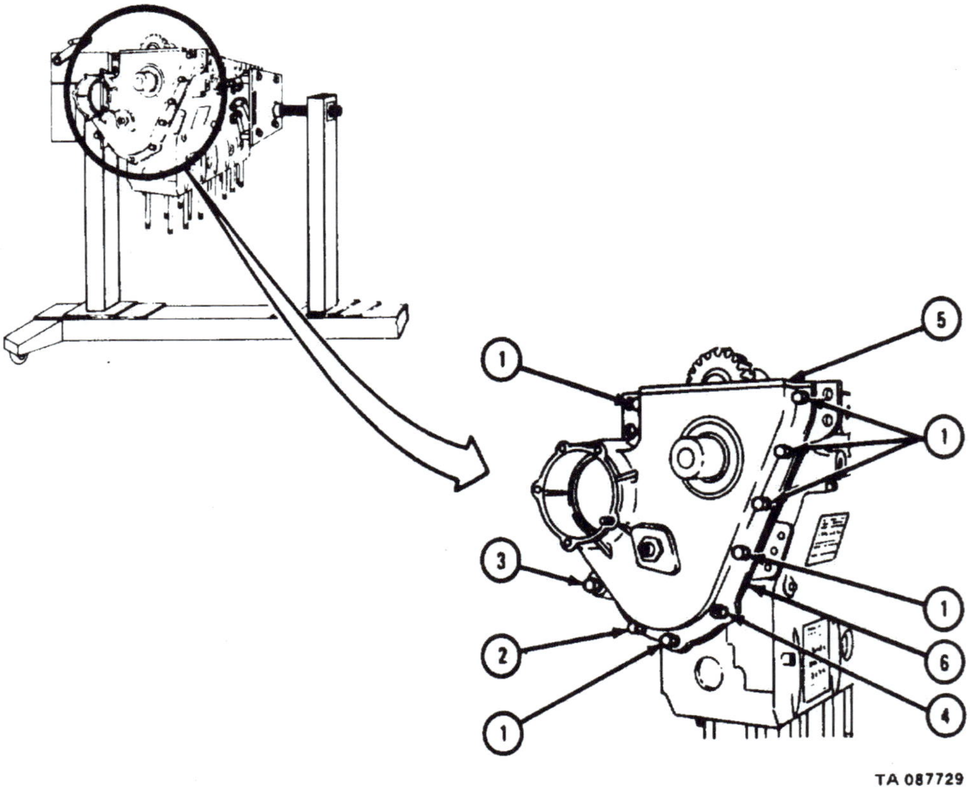

3-46. CAMSHAFT GEAR AND CRANKSHAFT GEAR.

FRAME 1

NOTE

Crankshaft dirt and liquid deflector (1) is used on engine LDS - 465-2 only.

1. Take of crandshaft dirt and liquid deflector (1).
2. Put wood block or brass rod (2) between camshaft gear (3) and crankshaft gear (4) as shown.
3. Take off camshaft gear retaining nut (5).
4. Take out wood block or brass rod.
5. Using mechanical puller, take out camshaft gear (3) and crankshaft gear (4).

END OF TASK

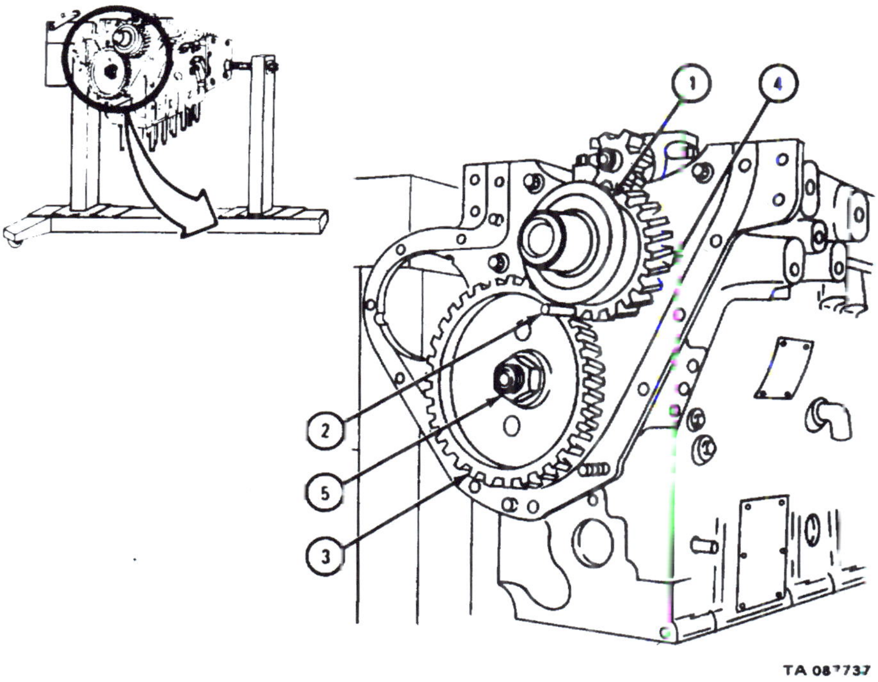

3-47. OIL PUMP IDLER GEAR ASSEMBLY.

FRAME 1

NOTE

Do not take off oil pump idler gear (3) if it is not damaged. If idler gear is not damaged, go to step 4.

1. Take off retaining ring (1).
2. Take off thrust washer (2).
3. Take off idler gear and bearing (3). Press bearing out of idler gear.

CAUTION

Be careful not to bend or scratch main bearing lower half (6) when taking it out.

4. Take out two screws and washers (4) and take out front main bearing cap (5). Take off front main bearing lower half (6).
5. Tag front main bearing lower half (6) with the number 1.

END OF TASK

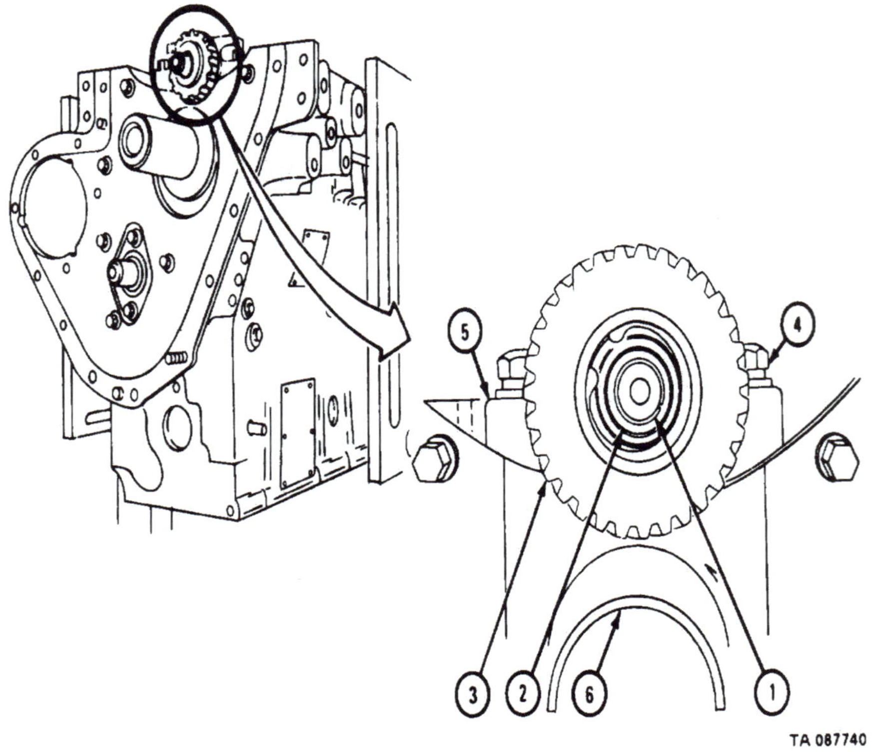

3-48. MAIN BEARING CAPS, CRANKSHAFT ASSEMBLY, AND MAIN BEARINGS.

CAUTION

It is easy to damage the equipment if you do not know what you are doing. Do not try to do this task unless you are experienced at it, or you have an experienced person with you.

FRAME 1

NOTE

Main bearing caps are called by numbers one to seven counting from front to rear of engine. The number four main bearing cap (6) is called thrust main bearing cap.

1. Using center punch, put two punch marks (1) on camshaft side of number two main bearing cap (2).
2. Using center punch, put three punch marks (3) on camshaft side of number three main bearing cap (4).
3. Using center punch, put four punch marks (5) on camshaft side of thrust main bearing cap (6).
4. Using center punch, put five punch marks (7) on camshaft side of number five main bearing cap (8).
5. Using center punch, put six punch marks (9) on camshaft side of number six main bearing cap (10).

NOTE

The number seven rear main bearing cap does not need punch marks because it can only go on one way.

GO TO FRAME 2

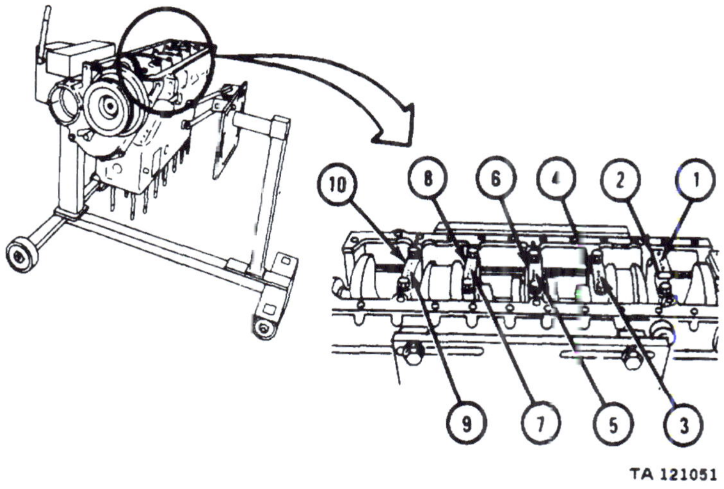

FRAME 2

CAUTION

Be careful not to bend or scratch main bearing lower half (3) when taking it out.

1. Take out two screws and washers (1) and take out rear main bearing cap (2). Take off rear main bearing lower half (3).
2. Tag rear main bearing lower half (3) with the number 7.
3. Take out and throw away two rear main bearing cap seals (4).

GO TO FRAME 3

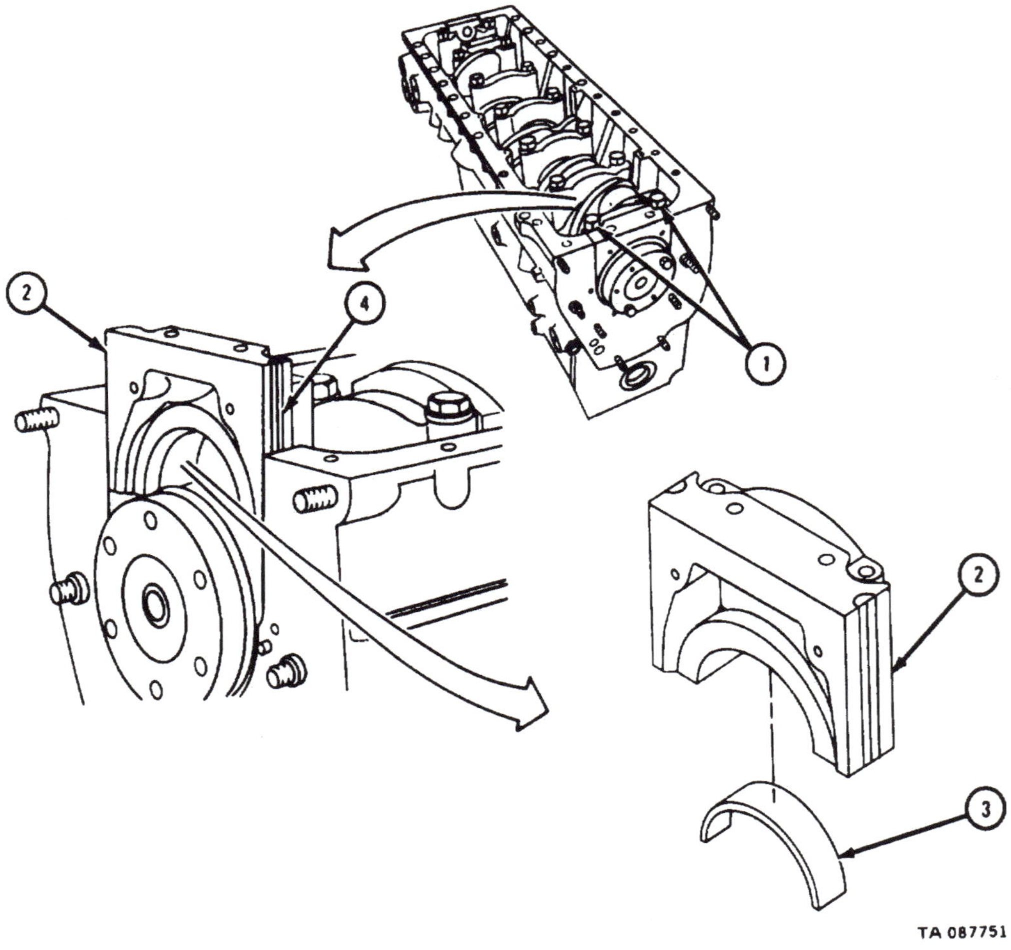

FRAME 3

CAUTION

Be careful not to bend or scratch main bearing lower half (3) when taking it out.

1. Take out two screws and washers (1) and take out front main bearing cap (2). Take off front main bearing lower half (3).

2. Tag front main bearing half (3) with the number 1.

GO TO FRAME 4

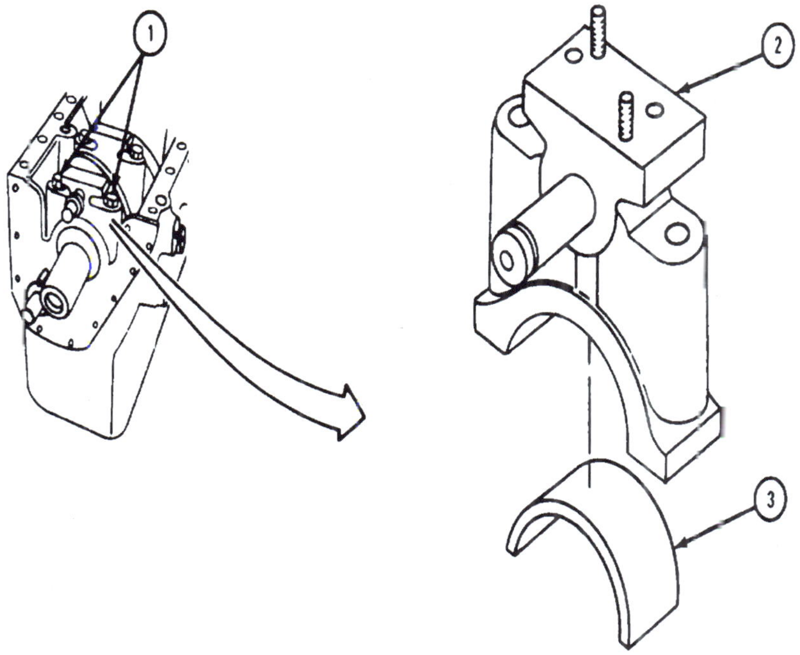

TA 087752

FRAME 4

CAUTION

Be careful not to bend or scratch main thrust bearing lower half (3) when taking it out.

NOTE

Main bearing caps and main bearing lower halves are called by numbers one to seven counting from front to rear of engine.

The number four main bearing cap (2) is called thrust main bearing cap. It is different-from other main bearing caps (4) because it has a thrust shoulder (5).

The number four main bearing lower half (3) is called thrust main bearing lower half. It is different from other main bearing lower halves because it has flanges (6).

1. Take out two screws and washers (1) and take out thrust main bearing cap (2). Take off thrust main bearing lower half (3).

GO TO FRAME 5

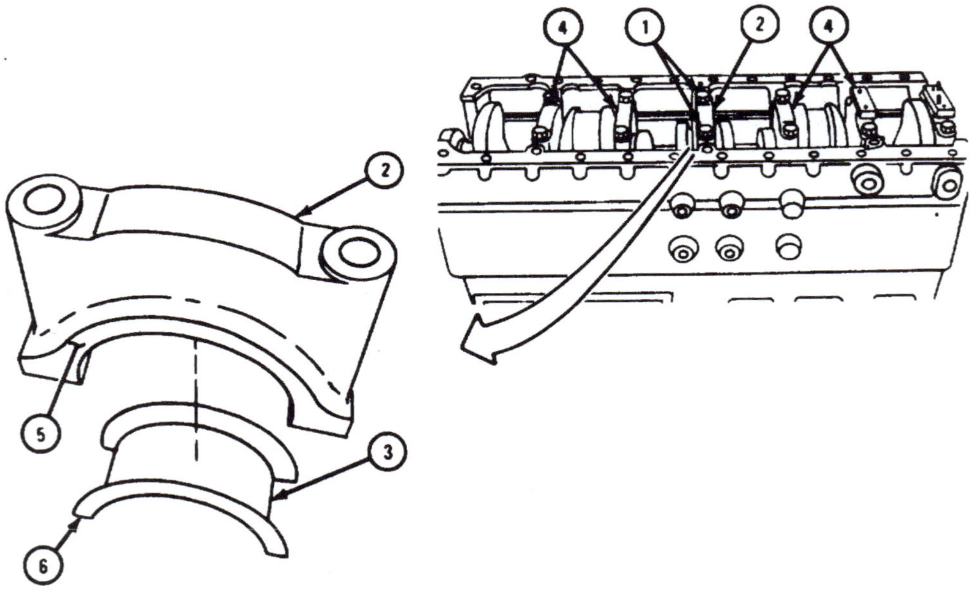

TA 113616

FRAME 5

CAUTION

Be careful not to bend or scratch main bearing lower halves (3) when taking them out.

NOTE

Main bearing caps (2, 4, 5, and 6) and main bearing lower halves (3) are called by numbers one to seven counting from front to rear of engine.

1. Take out two screws and washers (1) and take out number two main bearing cap (2). Take off number two main bearing lower half (3). Tag number two main bearing lower half with the number 2.
2. Do step 1 again for number three main bearing cap (4), number five main bearing cap (5), and number six main bearing cap (6).

GO TO FRAME 6

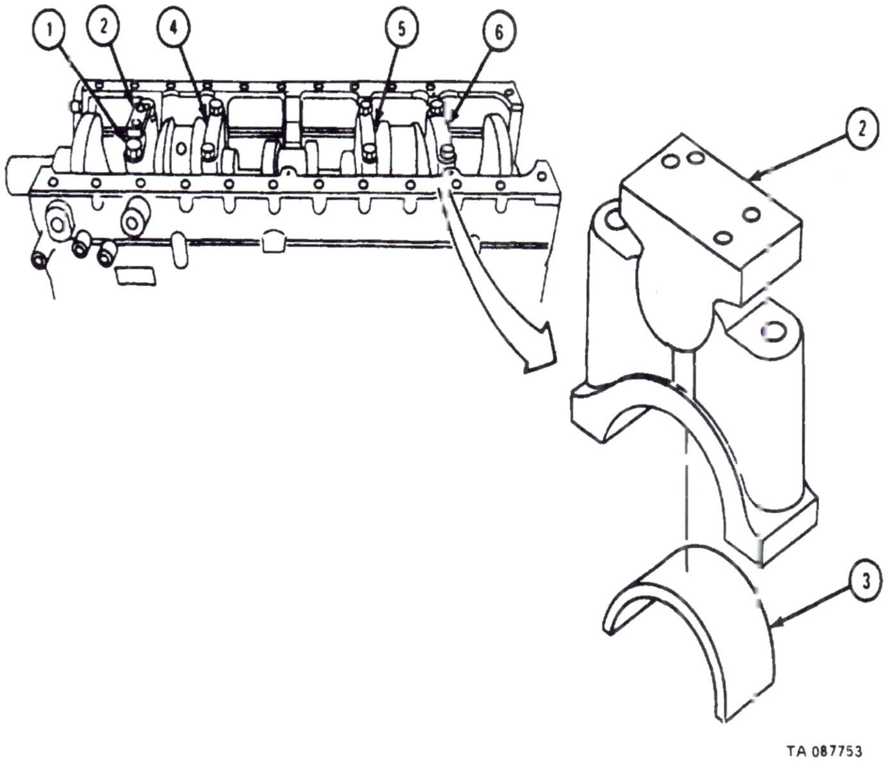

FRAME 6

CAUTION

Do not use a chain or metal cable sling to lift crankshaft (2). Crankshaft will be scratched.

1. Wrap rope sling (1) around crankshaft (2) as shown. Crankshaft must be balanced on sling so that it does not tip when lifted out.
2. Hook rope sling (1) to chain hoist (3).
3. Work chain hoist (3) to lift out crankshaft (2). Set crankshaft down on workbench.

GO TO FRAME 7

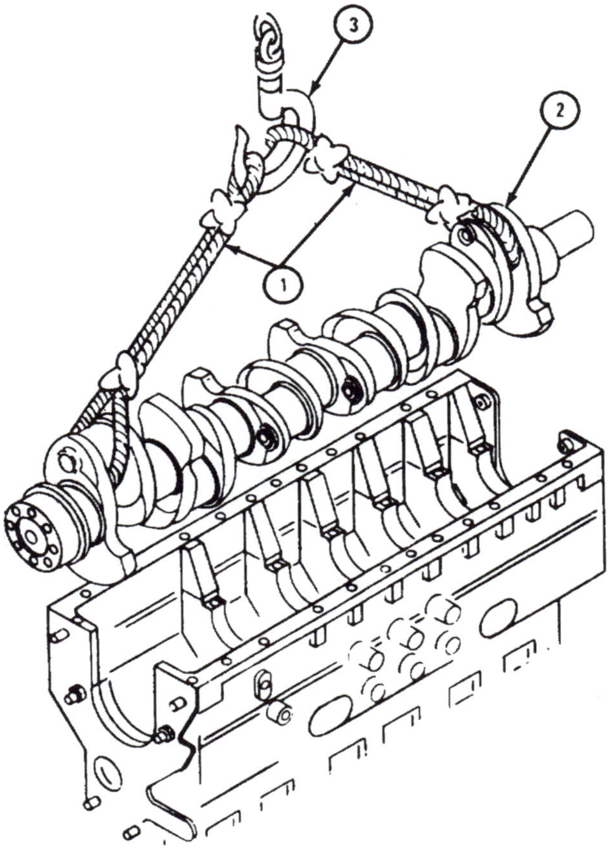

TA 113617

FRAME 7

CAUTION

Be careful not to bend or scratch main bearing upper halves (1 through 7) when taking them out

NOTE

Main bearing upper halves are called by numbers one to seven counting from front to rear of engine.

The number four main bearing upper half (1) is called main thrust bearing upper half. It is different from other main bearing upper halves (2 through 7) because it has flanges (8).

1. Take out main thrust bearing upper half (1).
2. Take out number one main bearing upper half (2). Using tool tag and wire, tag it with the number 1.
3. Do step 2 again for number two main bearing upper half (3), number three main bearing upper half (4), number five main bearing upper half (5), number six main bearing upper half (6), and number seven main bearing upper half (7).

END OF TASK

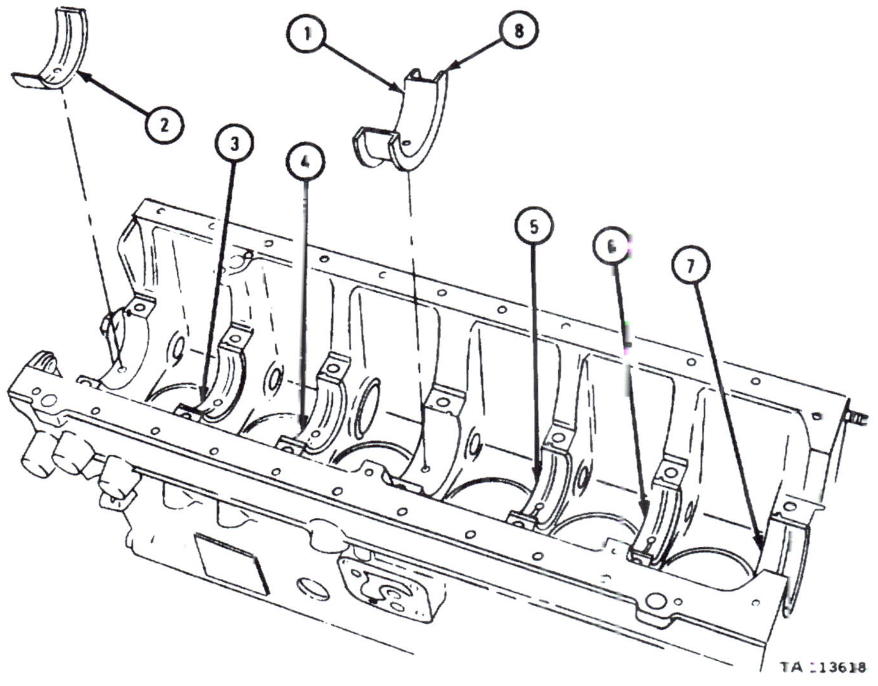

3-49. ENGINE FRONT PLATE.

FRAME 1

1. Takeout six screws and lockwashers (1) and takeoff engine front plate (2).
2. Take off and throw away engine front plate gasket (3).

END OF TASK

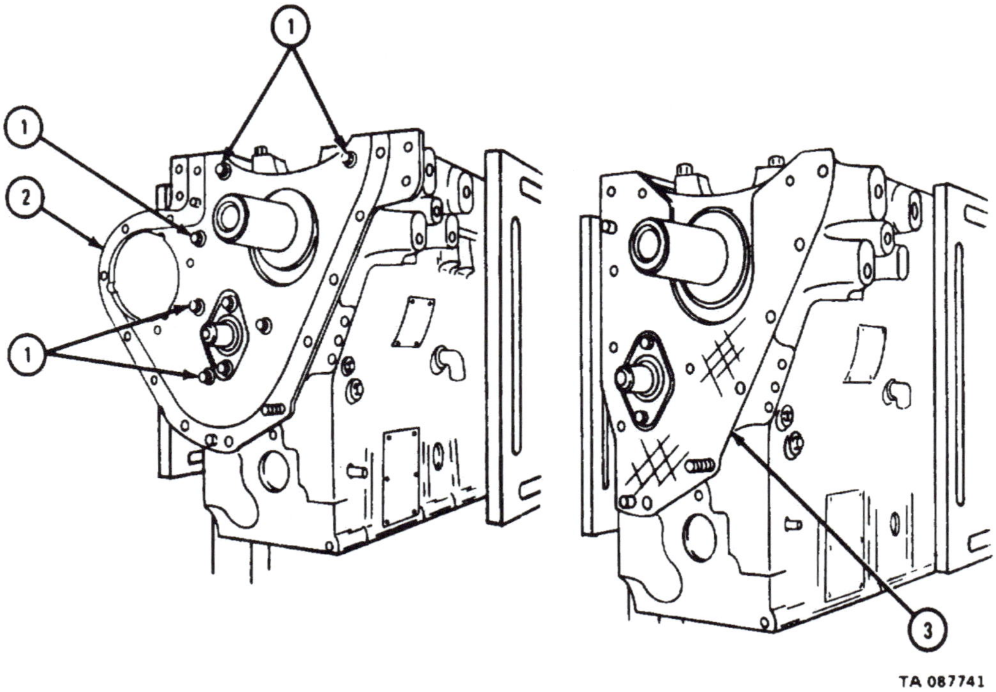

3-50. PISTON COOLING NOZZLES.

FRAME 1

NOTE

Cylinders are called by numbers one to six counting from front to rear of crankcase.

1. Take out six lockbolts (1) and six piston cooling nozzles (2). Tag each piston cooling nozzle with the number of cylinder it came from.

END OF TASK

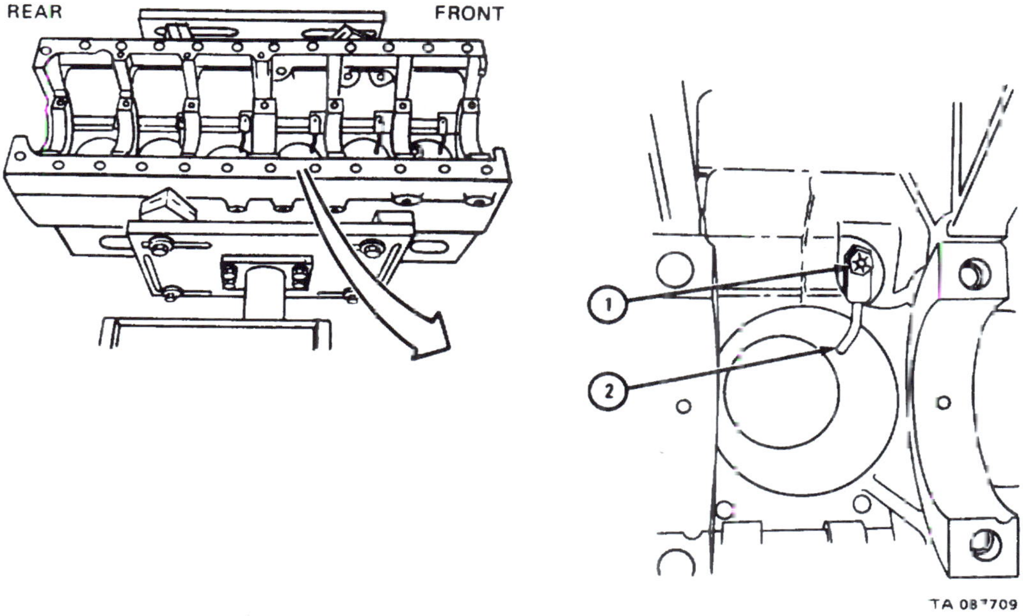

3-51. CAMSHAFT ASSEMBLY.

FRAME 1

1. Takeout two screws and washers (1) and take off camshaft thrust plate (2).

CAUTION

When taking out camshaft (3), be careful not to damage camshaft bearing.

2. Take out camshaft (3).

GO TO FRAME 2

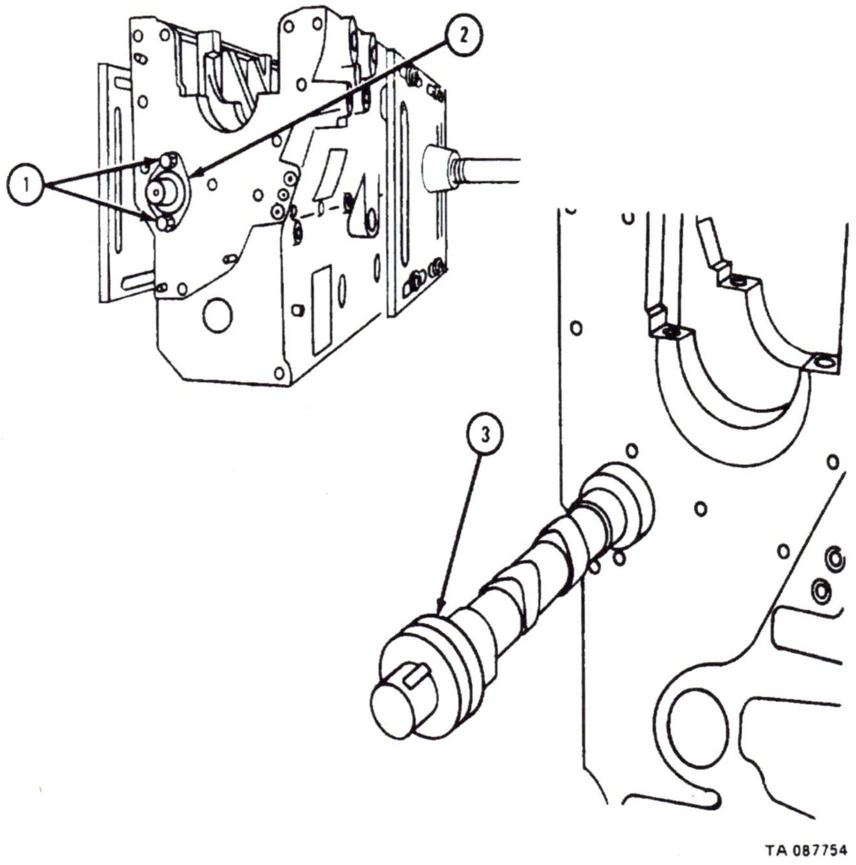

FRAME 2

CAUTION

When sliding long bar through camshaft bearings, be careful not to damage bearings.

1. Using long bar and soft-faced hammer, knock camshaft rear bearing plug (1) out by pounding inward from back of crankcase (2).

END OF TASK

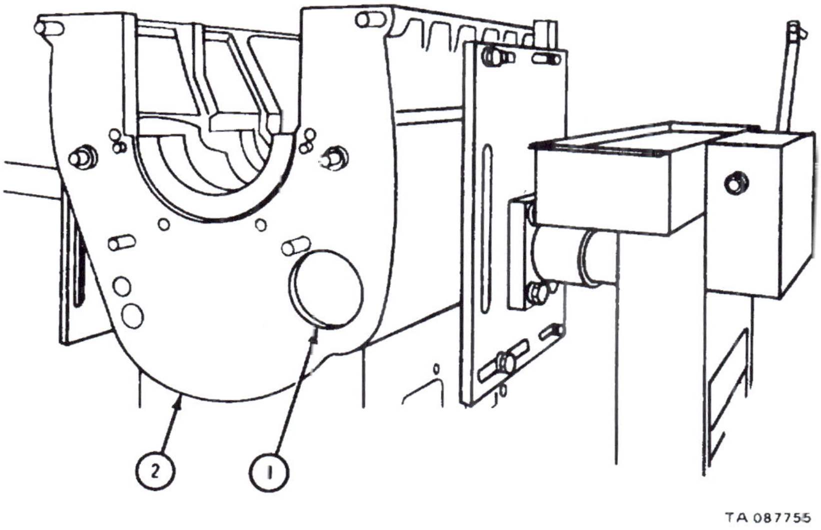

3-52. REMOVING CYLINDER BLOCK FROM OVERHAUL STAND.

FRAME 1

1. Put two crossbars (1) on four studs (2) and screw on four nuts (3).
2. Put lifting sling (4) with hoist hook (5) under two cross bars (1) as shown.
3. Raise hoist until lifting sling (4) is holding weight of cylinder block (6).
4. Working on side of overhaul stand (7) without crank handle, loosen three capscrews 8).
5. Take out three capscrews (9) and slide three universal mounting arms (10) away from cylinder block (6).

GO TO FRAME 2

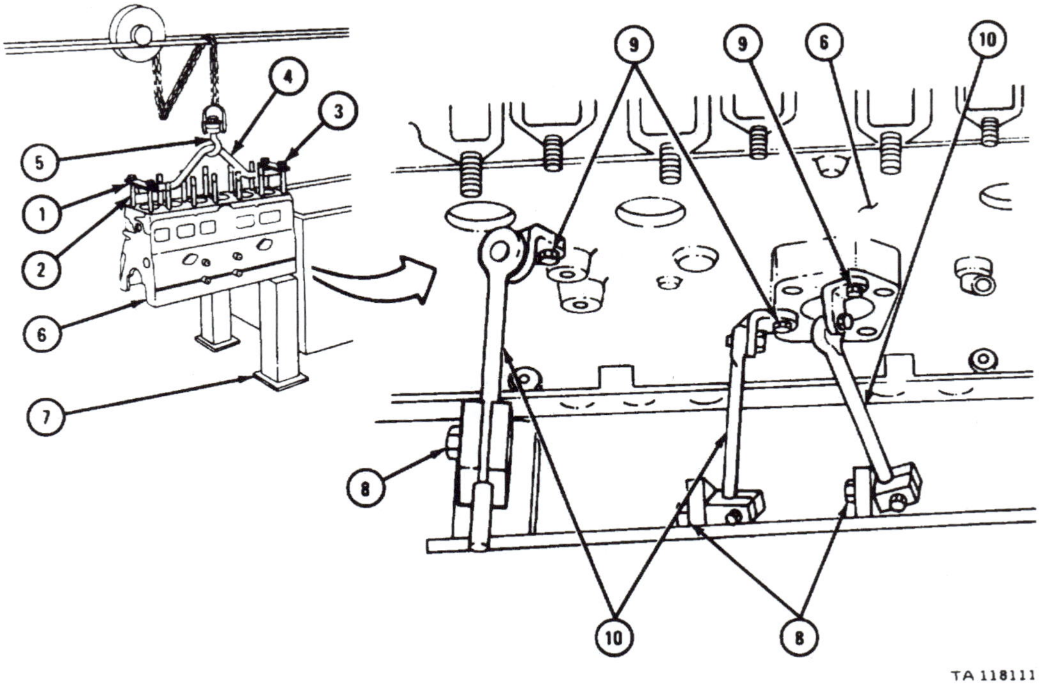

FRAME 2

1. Working on crank handle side, loosen three capscrews (1).
2. Take out three capscrews (2)
3. Takeout three capscrews (1) and three universal mounting arms (3).

GO TO FRAME 3

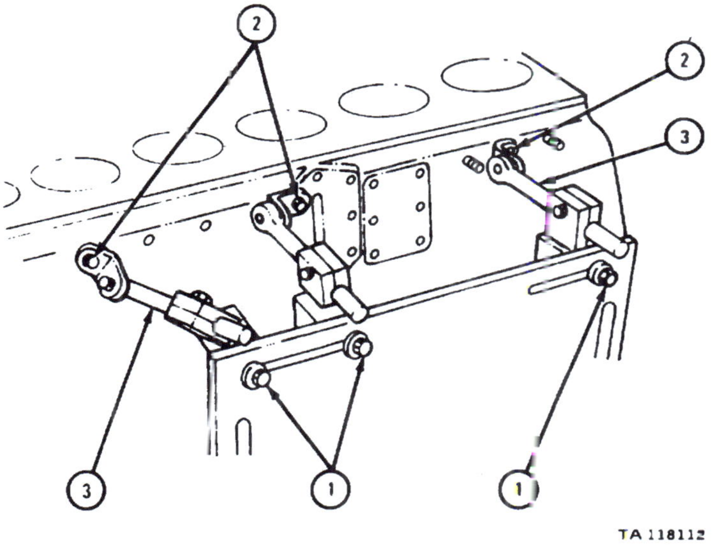

FRAME 3

1. Working on crank handle side, loosen two screws (1).
2. Take out two capscrews (2).
3. Takeout two capscrews (1).
4. Take off two universal mounting arms (3).

GO TO FRAME 4

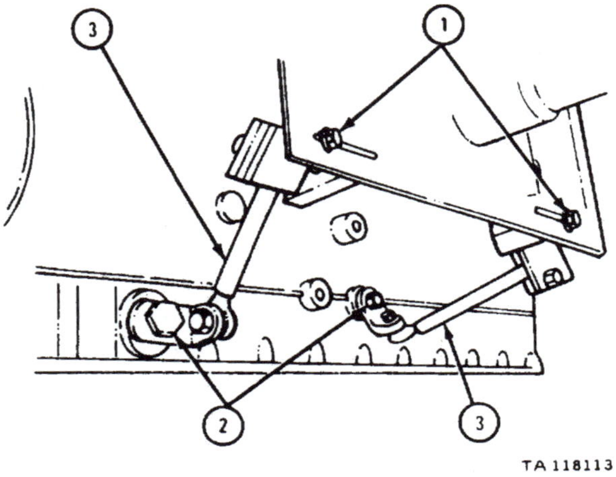

FRAME 4

1. Using chain hoist (1), lift cylinder block (2) clear of overhaul stand (3)
2. Move overhaul stand (3) out of the way .
3. Using chain hoist (1), lower cylinder block (2) down onto ground .

END OF TASK

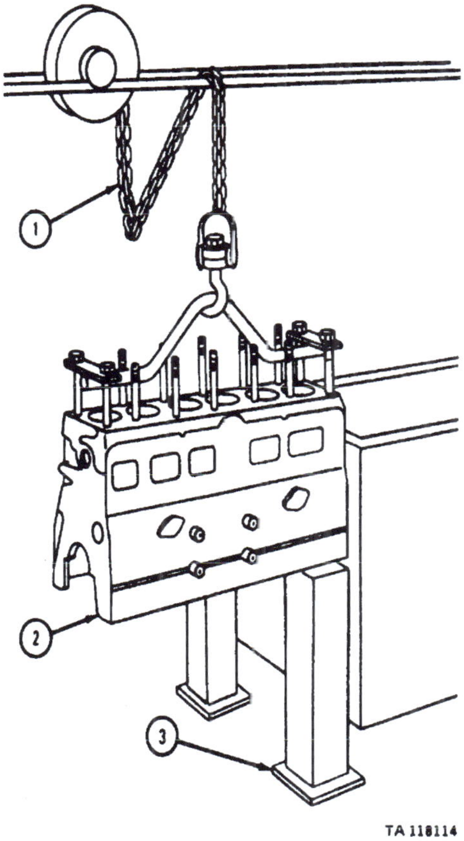

By Order of the Secretaries of the Army and the Air Force:

E. C. MEYER
General, United States Army
Chief of Staff

Official:

J. C. PENNINGTO N
Major General, United States Army
The Adjutant General

LEW ALLEN, JR., General, USAF
Chief of Staff

Official:

VAN L. CRAWFORD, JR., Colonel, USA F
Director of Administration

Distribution:

To be distributed in accordance with DA Form 12-38, Direct and General Support Maintenance requirements for 2-1/2 Ton Truck Cargo; 2-1/2 Ton Truck Van; 5-Ton Truck Chassis; 5-Ton Truck Van; and 5-Ton Truck Cargo; etc.

*U.S. GOVERNMENT PRINTING OFFICE: 1993-342-421/80780

THE METRIC SYSTEM AND EQUIVALENTS

LINEAR MEASURE

1 Centimeter = 10 Millimeters = 0.01 Meters = 0.3937 Inches
1 Meter = 100 Centimeters = 1000 Millimeters = 39.37 Inches
1 Kilometer = 1000 Meters = 0.621 Miles

WEIGHTS

1 Gram = 0.001 Kilograms = 1000 Milligrams = 0.035 Ounces
1 Kilogram = 1000 Grams = 2.2 Lb
1 Metric Ton = 1000 Kilograms = 1 Megagram = 1.1 Short Tons

LIQUID MEASURE

1 Milliliter = 0.001 Liters = 0.0338 Fluid Ounces
1 Liter = 1000 Milliliters = 33.82 Fluid Ounces

SQUARE MEASURE

1 Sq Centimeter = 100 Sq Millimeters = 0.155 Sq Inches
1 Sq Meter = 10,000 Sq Centimeters = 10.76 Sq Feet
1 Sq Kilometer = 1,000,000 Sq Meters = 0.386 Sq Miles

CUBIC MEASURE

1 Cu Centimeter = 1000 Cu Millimeters = 0.06 Cu Inches
1 Cu Meter = 1,000,000 Cu Centimeters = 35.31 Cu Feet

TEMPERATURE

$5/9\,(^{\circ}F - 32) = {}^{\circ}C$
212° Fahrenheit is equivalent to 100° Celsius
90° Fahrenheit is equivalent to 32.2° Celsius
32° Fahrenheit is equivalent to 0° Celsius
$9/5\, C^{\circ} + 32 = F^{\circ}$

APPROXIMATE CONVERSION FACTORS

TO CHANGE	TO	MULTIPLY BY
Inches	Centimeters	2.540
Feet	Meters	0.305
Yards	Meters	0.914
Miles	Kilometers	1.609
Square Inches	Square Centimeters	6.451
Square Feet	Square Meters	0.093
Square Yards	Square Meters	0.836
Square Miles	Square Kilometers	2.590
Acres	Square Hectometers	0.405
Cubic Feet	Cubic Meters	0.028
Cubic Yards	Cubic Meters	0.765
Fluid Ounces	Milliliters	29.573
Pints	Liters	0.473
Quarts	Liters	0.946
Gallons	Liters	3.785
Ounces	Grams	28.349
Pounds	Kilograms	0.454
Short Tons	Metric Tons	0.907
Pound-Feet	Newton-Meters	1.356
Pounds per Square Inch	Kilopascals	6.895
Miles per Gallon	Kilometers per Liter	0.425
Miles per Hour	Kilometers per Hour	1.609

TO CHANGE	TO	MULTIPLY BY
Centimeters	Inches	0.394
Meters	Feet	3.280
Meters	Yards	1.094
Kilometers	Miles	0.621
Square Centimeters	Square Inches	0.155
Square Meters	Square Feet	10.764
Square Meters	Square Yards	1.196
Square Kilometers	Square Miles	0.386
Square Hectometers	Acres	2.471
Cubic Meters	Cubic Feet	35.315
Cubic Meters	Cubic Yards	1.308
Milliliters	Fluid Ounces	0.034
Liters	Pints	2.113
Liters	Quarts	1.057
Liters	Gallons	0.264
Grams	Ounces	0.035
Kilograms	Pounds	2.205
Metric Tons	Short Tons	1.102
Newton-Meters	Pound-Feet	0.738
Kilopascals	Pounds per Square Inch	0.145
Kilometers per Liter	Miles per Gallon	2.354
Kilometers per Hour	Miles per Hour	0.621

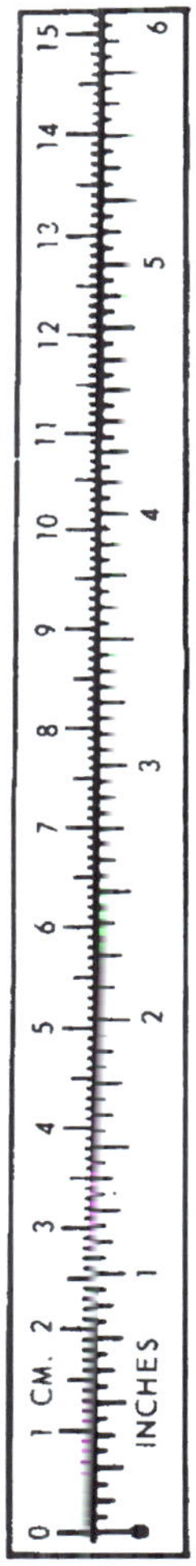

TA089991

PIN:048857-00 0

Made in United States
Troutdale, OR
11/03/2026

56844653R00144